THE SUTTER BUTTES
OF CALIFORNIA

THE SUTTER BUTTES
OF CALIFORNIA

A Study of Plio-Pleistocene Volcanism

BY

HOWEL WILLIAMS and G. H. CURTIS

UNIVERSITY OF CALIFORNIA PRESS
BERKELEY • LOS ANGELES • LONDON

UNIVERSITY OF CALIFORNIA PUBLICATIONS IN GEOLOGICAL SCIENCES

Editorial Board: D. I. Axelrod, W. B. N. Berry, R. L. Hay, M. A. Murphy,
J. W. Schopf, M. O. Woodburne, Chairman

Volume 116
Approved for publication March 26, 1976
Issued January 15, 1977

University of California Press
Berkeley and Los Angeles
California

University of California Press, Ltd.
London, England

ISBN: 0-520-09559-6
Library of Congress Catalog Card Number: 76-14296

Contents

Abstract 1
Introduction and Acknowledgments 2
Plutonic-metamorphic Basement 5
Upturned Cretaceous and Tertiary Beds 9
Pre-explosive Uplift of Sedimentary Beds 14
Extrusive and Intrusive Domes 20
Rampart Beds 26
Central Lake-beds 35
Intra-volcanic Deformation 37
The Buried "Colusa Buttes" 40
Age of Volcanism 41
Appendix: Petrography 44
Petrology 49
References 55
Plates 57

THE SUTTER BUTTES OF CALIFORNIA
A Study of Plio-Pleistocene Volcanism

by

HOWEL WILLIAMS AND G. H. CURTIS

Abstract

The Sutter (Marysville) Buttes rise in dramatic isolation from the wide floor of the Sacramento Valley. They occupy a circular area 10 miles across, made up of four distinctive topographic and geologic units: (1) a peripheral ring, *"The Rampart,"* which rises inward gently and consists almost wholly of fluviatile, volcanic deposits; (2) an inner ring, *"The Moat,"* formed of gently rounded hills of upturned Cretaceous and Tertiary sediments; (3) a craggy, *"Castellated Core,"* occupied mainly by a cluster of Pelean domes, the most conspicuous of which are North Butte and South Butte, rising to heights of 1,863 and 2,132 feet, respectively; (4) *Central Lake-beds.*

The plutonic-metamorphic basement is exposed at the edge of the Sierra Nevada at elevations of about 100 feet; thence, it falls westward until it lies roughly 6,600 to 7,900 feet beneath the Sutter Buttes where it consists partly of norite but chiefly of quartz diorite of late Jurassic age. West of the Buttes, the basement descends abruptly, until, 20 miles away, along the edge of the Coast Ranges, it lies at depths of 35,000 to 40,000 feet, consisting primarily of Franciscan metamorphic rocks. A narrow belt of magnetic and gravity "highs" runs lengthwise along the center of the San Joaquin-Sacramento Valley, passing close to the western side of the Sutter Buttes, roughly where the "Bedrock surface" begins its sharp descent.

There are no lower Cretaceous beds in the Sutter Buttes, although in the Coast Ranges, 20 miles to the west, these are approximately 16,000 to 20,000 feet thick. Upper Cretaceous beds in the Buttes reach a maximum thickness of about 5,000 feet. They consist principaly of shales, but near their top include gas-rich, white Kione sands, formerly called Ione and wrongly considered to be of Eocene age. Eocene beds lie unconformably on Cretaceous sediments in the following order: Capay Shales; Ione Sands; and Butte Gravels, marine equivalents of the fluviatile Auriferous Gravels of the Sierra Nevada. Conformably above the foregoing lie Sutter Beds, mostly continental and deltaic, volcaniclastic sediments. Near their base, they contain thin lenses of airborne rhyolitic ash, equivalents of some of the rhyolitic tuffs in the Valley Springs Formation of the Sierra Nevada, known to range in age from Late Oligocene to early Miocene. Most of the Sutter Formation, however, consists of Mio-Pliocene andesitic sands and silts, outwash equivalents of the voluminous laharic deposits of the Mehrten formation of the Sierra Nevada.

When regional uparching of the Cretaceous and Tertiary beds began in the Sutter Buttes and how long it lasted before the first explosions took place cannot be told. Uplift was produced, not by a simple andesitic laccolith, as formerly supposed, but by a group of intrusions probably composed chiefly of rhyolite with subordinate dacite and andesite. The uparched beds were broken into sectors that were tilted outward at various angles. On the south side, sediments approximately 5,000 feet thick were upturned; on the opposite side, a thinner pile of sediments was tilted outward at angles of only 10° to 20°. A few roughly concentric faults may have accompanied the regional uplift. How high the initial uplift rose is also uncertain, but its rise must have been attended by rapid erosion of the sedimentary cover, and by repeated landslides on the flanks. Where uplift was greatest, the roof may have risen to heights of approximately 2,500 feet; elsewhere, it was only about 2,000 feet high, or even lower, and probably there was an east-west depression in the sedimentary cover, roughly on the present site of Bragg's Canyon.

Many domes surmount and surround the initial intrusions, some extrusive and others intrusive at depth. These range in composition from andesite to rhyolite, though they were not emplaced in any regular order. Some of the oldest domes rose in the sedimentary moat; some of the youngest formed conspicuous peaks above the initial stocks in the center of the buttes during and after the final stages of explosive activity, e.g. North, South, and West Buttes. A few rhyolite and dacite domes rose within the interior of the buttes, but most rose within the moat or still farther away, beneath the encircling rampart, and even beyond. A few of these siliceous domes may have been emplaced before explosive activity began; more likely all rose during the explosive phase, and hence uparched not only the sedimentary beds but also the volcaniclastic Rampart beds.

Extrusive domes are all of Pelean type. Most are circular, measuring between a quarter and half mile across. Flow-banding is most pronounced among rhyolitic domes where it is generally concentric with the margins and is either vertical or dips inward at high angles. Autobrecciation is also most pronounced in rhyolitic domes, some of which are bordered by *peperitic aureoles*, caused by steam-blast explosions when viscous magma entered water-soaked sediments near the surface. All domes, whether extrusive or still deeply buried, strongly uparched adjacent beds but none produced more than slight induration of immediately adjacent wall-rocks.

Explosive volcanism did not begin until the sedimentary beds in the moat had been eroded almost to their present levels. Explosions took place from a cluster of vents, but instead of building a single huge cone, as formerly envisaged, they produced a swarm of relatively small cones, none of which rose to heights of much more than 2,500 feet above sea-level. Explosions may have recurred at short intervals for a million years, and were accompanied by the intrusion and extrusion of many domes.

A few explosions, chiefly the early ones, involved discharge of fresh magma in the form of pumiceous ejecta, but most were weak steam blasts (phreatic eruptions) involving discharge of lithic fragments that were already solid when expelled. No large bombs or blocks were blown out; indeed, few ejecta exceeded the sizes of sand and gravel. The original cones and conelets have long since been removed by erosion. Some ejecta were washed into a large central lake-basin, but most were carried outward by streams to build an enormous volcaniclastic cone, the remnants of which now form the peripheral rampart of the buttes.

Three members may be crudely distinguished within the Rampart formation, as follows: a basal member composed of pale-colored, waterlaid tuffaceous sediments derived from vitric and pumiceous, magmatic ejecta composed of rhyolite, dacite, and siliceous hornblende-biotite andesite, admixed with a slight amount of pre-volcanic, sedimentary debris; a middle member, by far the thickest, composed mainly of reworked andesitic debris; and an upper member, composed almost wholly of bouldery andesitic material laid down by torrential lahars that debouched from the steep flanks of the youngest Pelean domes in the interior of the buttes (e.g. North, South, and West Buttes).

A small flattish cone of andesitic lava—the De Witt volcano—was finally built on top of the Rampart beds near the southeast base of the buttes. The total volume of airfall and waterlaid ejecta deposited during the long explosive phase was probably less than 2 cubic miles.

Potassium-argon measurements suggest that the basal Rampart beds are approximately 2.5 m.y. old, and andesitic lava erupted by the De Witt volcano after explosive activity ended, is about 1.5 m.y. old.

While volcanism was going on in the Sutter Buttes, viscous intrusions of rhyolite were emplaced roughly 4 miles to the west, deep beneath the floor of the Sacramento Valley. Some of these intrusions were cut by gas wells at depths of between 5,600 and 9,470 feet. Together, they produced an oval-shaped uplift called the "Colusa Buttes," measuring 12 by 3 to 4 miles across, in which some Cretaceous beds were uparched more than 1,500 feet. These deeply buried rhyolites have crypto- and micro-crystalline matrices indistinguishable from those of rhyolites extruded in the Sutter Buttes. One sample was determined to be 2.2 m.y. old.

The igneous rocks of the Sutter Buttes are mainly porphyritic hornblende-biotite andesites, hornblende-poor biotite dacites and biotite rhyolites belonging to the calc-alkaline igneous series. Their alkali-lime index approximates 60. They contain more potash and less lime than coeval lavas in the High Cascades with the same silica-content. Judging by the potash content of andesites containing 60% SiO_2, the parent magmas that fed the High Cascade volcanoes rose from a subduction-zone at a depth of 110 to 140 km; on the same basis, andesites from the Newberry volcano of Central Oregon, and the Medicine Lake Highlands of Northern California, which lie east of the Cascades, being richer in K_2O, are thought to have come from depths of 150 to 160 km. However, the Sutter Buttes andesites, though they lie west of the southern end of the Cascades, are still richer in K_2O and hence seem to have originated at even greater depths (approx. 180 km).

The Sutter Buttes lavas are notably richer in Ba and Sr than most circum-Pacific coeval lavas of the same silica-content.

Introduction and Acknowledgments

The volcanoes that form the Sutter Buttes, formerly called the Marysville Buttes, rise in spectacular fashion from the wide, flat floor of the Sacramento Valley (fig. 1). Their remarkable isolation immediately arrests attention. No other volcanoes with which we are familiar are associated with greater deformation of surrounding sedimentary beds. Pro-

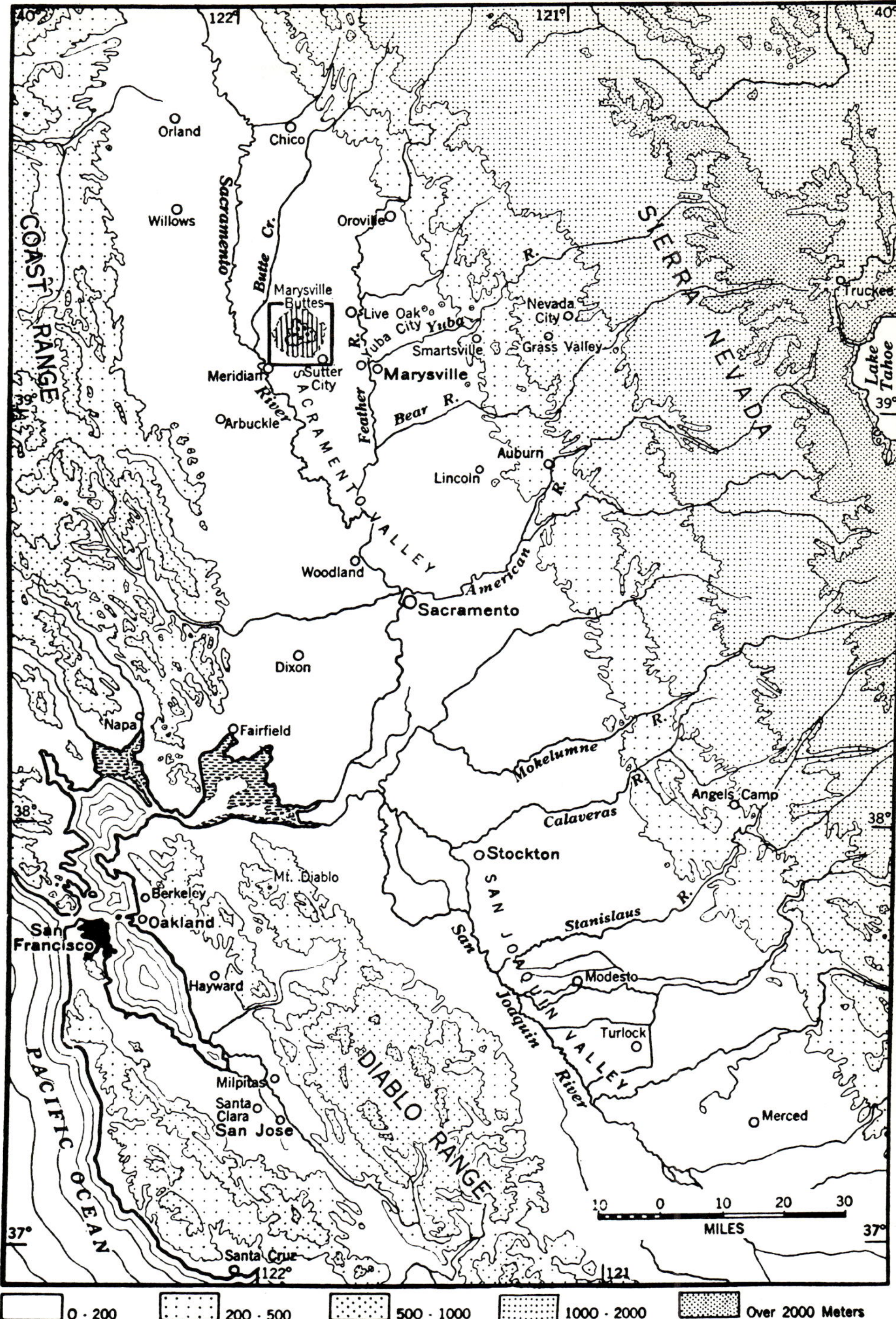

Fig. 1. Index map of a portion of California showing the Sutter (Marysville) Buttes in relation to the Sacramento Valley, Sierra Nevada, and Coast Ranges.

nounced and widespread uplifts accompanied the rise of central igneous stocks in the Henry and La Sal Mountains of Utah in mid-Tertiary time, and localized doming accompanied emplacement of peripheral laccoliths, but there was no explosive activity in the Henry Mts. and very little in the La Sal Mts. compared with the long-continued and large-scale volcanism in the Sutter Buttes. In the Sutter Buttes, a large central uplift of sedimentary beds, locally more than 5,000 feet thick, was followed by explosive activity that continued intermittently for at least half a million years, and by many peripheral intrusions and extrusions, each of which strongly uparched the surrounding sediments.

Eighty years have passed since Lindgren made the original study of the area, and though he spent no more than a few weeks there, with only a crude topographic map, he correctly envisaged the principal volcanic events. Subsequently, Dickerson (1913) made valuable contributions to knowledge of the Cretaceous and Tertiary beds upturned during the volcanic episode. Then, in 1926, one of us (Williams), fresh from studying Ordovician volcanic rocks in North Wales and eager to familiarize himself with younger volcanic fields, was urged by the late Professor G. D. Louderback to examine the Sutter Buttes. Few aerial photographs of the region were available in those days, no gas wells had been drilled, no geophysical surveys had been made, and of course radiometric dating by the potassium-argon method had still to be invented! A vast amount of new geological and geophysical information has now been gathered and the time seems ripe for a radical modification of earlier views.

Our interest in the volcanic history of the Sutter Buttes was revived by events that took place in 1944-1945 at Syowa Sinzan in Hokkaido, Japan. Flat ground near the foot of Usu volcano was slowly upheaved by a rising body of dacite, and during the closing stages of uplift, steam-blast explosions formed a cluster of small craters on top of the rising "Roof Mountain." Finally, a steep-sided Pelean dome of dacite rose through the uplifted beds. What happened during this brief episode took place at the Sutter Buttes on a vastly larger scale and over a vastly longer period. Our interest in the buttes was also revived by the wealth of new information provided by drilling and geophysical exploration. Accordingly, in 1951 we began to re-map the area, making many brief visits at long intervals, sometimes singly, sometimes together, and occasionally with classes of students. Curtis concentrated his mapping on the southern part of the buttes, discovering the De Witt volcano, and he is responsible for radiometric dating of many of the rocks; Williams assisted in re-mapping the southern part and emended his mapping of the remainder of the buttes. He is also responsible for writing the text and preparing the illustrations.

Geologists of several oil companies gave welcome aid. Not only did they give valuable stratigraphic and geophysical information, but they also assisted by obtaining samples of deeply buried "bedrocks" beneath the buttes and adjacent parts of the Sacramento Valley. Their contributions to the geologic history of this part of California merit warm appreciation. Particular thanks go to W. R. "Sam" Brown, of the Buttes Gas and Oil Company; to Douglas Thamer, formerly with the same company; and to Ernest McKittrick and Walt Smith, formerly with the Shell Oil Company. It is also a pleasure to acknowledge helpful discussions with our Berkeley colleagues, Fred Berry and Clyde Wahrhaftig. Trace-element determinations were kindly made by Bob Jack; Robert Drake made several K/A age-determinations, and Joaquim Hampel was generous in sharing his skills in photography. Ernest Wettstein of Yuba College facilitated field work in many ways.

Our main concern is to discuss the volcanic history of the Sutter Buttes. We have considered it proper, however, to review the pre-volcanic history, partly to correct some of Williams' early errors and partly to present new information concerning the plutonic-metamorphic basement beneath and around the buttes.

The aerial photograph reproduced as plate 1 depicts in dramatic fashion the isolation of the buttes within the Sacramento Valley. They occupy a circular area approximately 10 miles in diameter which, for convenience in description, can be divided into the following four units.

1. *The Rampart.* A peripheral belt consisting almost entirely of waterlaid volcanic deposits formed by reworking of airfall-ejecta blown from vents in the interior of the buttes. The gentle outer slopes of the rampart rise to a crest that surrounds the second topographic unit.

2. *The Moat.* An annular belt typified by round-topped hills and ridges eroded in upturned Cretaceous and Tertiary sedimentary beds. Within the moat rise several craggy knolls of white rhyolite and a few darker ones of andesite, the relics of Pelean domes.

3. *The Castellated Core.* Most of the interior of the buttes is occupied by a coalescing cluster of Pelean domes of andesite, the three most conspicuous of which are South Butte (2,117 feet), North Butte (1,863 feet), and West Butte (1,681 feet).

4. *The Central Lake-beds.* A thick pile composed mainly of lacustrine, volcanic sediments surrounded and partly deformed by andesitic domes.

The foregoing fourfold division of the Sutter Buttes, to which repeated reference is made in the sequel, is depicted in figure 2 and plates 1, 2, and 3.

Fig. 2. Sutter Buttes viewed from the west. Outward-dipping slopes of the Rampart conceal the sedimentary moat. The rugged core of the buttes is made up chiefly of Pelean-type domes of andesite.

Plutonic-Metamorphic Basement

Scores of wells have been drilled in the Sacramento Valley in search of gas and oil, but only a few have penetrated the Cretaceous and Tertiary sedimentary beds to cut the underlying "basement," and none penetrated this for more than a few feet. Nevertheless, available well-data, coupled with geophysical information, enable us to give a general account of the topography and lithology of the basement under the Sutter Buttes and adjacent parts of the valley.

Depth of the Basement

The "bedrock surface" falls gently from the edge of the Sierra Nevada to midway across the Sacramento Valley, to the Sutter Buttes and vicinity, and thence falls abruptly toward the Coast Ranges. At the edge of the Sierra, bedrocks outcrop at elevation of about 100 feet above sea-level; beneath the Sutter Buttes, they lie at depths of approximately 6,500 to 7,900 feet. A few miles to the south and west, they drop to depths of more than 10,000 feet.

Along the west side of the Sacramento Valley, only 20 miles from the Sutter Buttes, the basement of metamorphic Franciscan rocks lies at depths of 35,000 to 40,000 feet (Lachenbruch, 1962, fig. 2). Correspondingly, Cretaceous sediments of the "Great Valley sequence" thicken from a feather-edge at the foot of the Sierra Nevada to approximately 3,000 feet within the Sutter Buttes, and thence to more than 30,000 feet along the west side of the Sacramento Valley.

A prominent magnetic high, closely associated with a gravity high, runs along the center of the Sacramento-San Joaquin Valley from Red Bluff to Bakersfield (Griscom, 1966). The axis of this belt passes close to the western edge of the Sutter Buttes and approximately through the buried "Colusa Buttes." Its origin is still in doubt. Grantz and Zeitz (1960) attributed it to the structure and lithology of the overlying sedimentary beds, whereas Bowers (1958; quoted by Chapman, 1966) attributed it to a buried ridge of basement rocks. There is, however, no convincing evidence for either opinion. Griscom suggested that the magnetic highs might be related to deeply buried mafic and ultramafic intrusions. Magnetic highs near Bakersfield are known through drilling to be associated with gabbros, serpentinites, and mafic volcanic rocks; in the Sierran foothills such highs seem to be related, not to metavolcanic rocks but to serpentinites and mafic intrusions.

Not enough wells have penetrated the basement beneath the Sutter and Colusa Buttes to provide a satisfactory explanation of the geophysical anomalies. See Nature of the Basement Rocks, below, for available data. It must suffice to note here that except for a gabbro and a metabasalt or metadiabase cut at depths of approximately 10,000 feet beneath the Colusa Buttes and a norite cut at about 6,500 feet beneath the Sutter Buttes, virtually all of the other gas wells that penetrated the basement within and near the buttes cut diorites and quartz diorites.

Why the Sutter and Colusa Buttes are localized where they are, within and close to the north-south belt of magnetic and gravity highs, we cannot say. There are no structural or lithological features within either the Sierra Nevada or the Coast Ranges that trend toward these Plio-Pleistocene volcanic centers. Their position may be related to an approximately north-south belt within which the plutonic-metamorphic basement begins to deepen westward at an increasing rate.

It should be noted in conclusion that the plutonic-metamorphic basement beneath the Sutter Buttes lies at depths of 6,561 to 7,868 feet, implying a local relief of more than 1,300 feet. Eastward, the basement rises rapidly: only a mile east of the buttes, it rises to a depth of 4,562 feet; 8 miles to the east it rises to 1,439 feet. In the opposite direction, about 4 miles west-northwest of the buttes, the basement drops to 10,127 feet. It also drops rapidly to the south, falling more than 1,300 feet from the edge of the andesitic core to beneath the outer edge of the surrounding rampart.

Nature of the Basement-Rocks

Some of the data to be recorded have already been published by May and Hewitt (1948); others have been supplied to us by geologists of the Buttes Gas and Oil and Shell Oil companies.

Before describing basement-rocks cut by wells in the northern part of the Sacramento Valley, it seems advisable to note that at least three wells north of the Sutter Buttes cut what seem to be Tertiary and Quaternary intrusive rocks. Two wells near Red Bluff cut remarkably fresh, coarse-grained, ophitic olivine-augite diabase. One well is located about 10 miles north of Red Bluff (Sec. 36, T. 29 N., R. 4 W.). This cut diabase at depths between

5,161 and 5,168 feet. The other well (Sec. 17, T. 28 N., R. 3 W.) cut similar diabase at depths between 3,465 and 3,481 feet. (Thin-sections are stored in Univ. of Calif. Berkeley Coll. 134, Nos. 106 and 107.) These diabases may represent fillings of magma chambers that fed some of the pre-Tehama Tertiary basaltic flows cut by other wells in the northern part of the Sacramento Valley (Hackel, 1966).

A third well cut igneous rocks at a depth of 5,269 feet near Gridley, about 4 miles northeast of the Sutter Buttes (Sec. 4, T. 17 N., R. 2 E.). Almost surely the "biotite dacite porphyry" cut by this well is a Quaternary intrusion of the same age as some of the rhyolite domes within the Sutter Buttes.

Most basement-rocks cut by wells in the central and eastern parts of the Sacramento Valley are diorites and quartz diorites; a few are gabbros and epizone schists. Some of these are now described as if collected during a traverse from north to south. (See fig. 3.)

North of the Sutter Buttes

1. *Sec. 16, T. 20 N., R. 3 W.* Close to the north end of the Sacramento Valley; north of the Willows-Beehive Bend gas field. Depth, 12,593 feet. (Univ. Calif. Berkeley Coll. 134-109.) *Pyritized keratophyre.* Consists of oligoclase laths with intersertal chlorite and epidote; contains several quartz-filled amygdules. It is either a Sierran or Franciscan metavolcanic rock.

2. *Sec. 11, T. 19 N., R. 2 W.* Approximately 18 miles north-northwest of the Sutter Buttes. Depth, 10,807 feet. *"Diorite."*

3. *Sec. 14, T. 18 N., R. 1 E.* Humble-Elna Schor Well No. 2; approximately 7 miles north of the Sutter Buttes. Depth, 5,666 feet. Identified megascopically by others as *augite gabbro.*

4. *Sec. 17, T. 17 N., R. 1 E.* Humble Wild Goose Well No. 6; approximately 4 miles northwest of the Sutter Buttes. Depth, 6,697 feet. (Univ. Calif. Berkeley Coll. 134.) *Mylonitized hornblende-rich diorite.*

5. *Sec. 32, T. 17 N., R. 1 E.* Approximately 1 mile northwest of the Sutter Buttes. Depth, 8,416 feet. Described by others simply as "basement rock."

6. *Sec. 27, T. 17 N., R. 1 E.* Close to the northern edge of the Sutter Buttes, about 3 miles west of Pennington. Depth, 6,640 feet. Well cut either basement or Quaternary volcanic rocks.

7. *Sec. 3, T. 16 N., R. 1 W.* Humble Capital Co. B-1 Well. Within the buried "Colusa Buttes," approximately 6 miles west of the Sutter Buttes. Depth, 10,127 to 10,133 feet. Referred to erroneously as a metadacite in Calif. Div. Mines and Geol. Report, 1962. Thin sections show one sample to be a slightly *uralitized augite gabbro* and the other to be an *amygdaloidal metabasalt or metadiabase* consisting chiefly of albite, actinolite, and ilmenite, the amygdules being filled with quartz, biotite, chlorite, epidote, and clinozoisite.

8. *Sec. 13, T. 16 N., R. 2 E.* Approximately 2 miles east of the Sutter Buttes. Depth, 4,562 to 4,931 feet. *"Granitic rubble."*

Beneath the Sutter Buttes

1. *Sec. 12, T. 16 N., R. 1 E.* Close to the northern edge of Peace Valley. Shell-Buttes Community No. 1 Well. At depths of 6,561 to 6,565 feet, cut *norite.* Half of the rock consists of fresh laths of labradorite, a third of augite, a tenth of hypersthene, and the remainder of granular iron ore. The minerals form a mosaic of interlocking, roughly equant grains most of which measure between 0.5 and 1.0 mm. across. A K/A test yielded an *age of 136 m.y.* (Late Jurassic or Early Cretaceous.) This well also cut several so-called sills of Quaternary rhyolite separated by Cretaceous shales at depths of about 6,000 feet.

2. *Sec. 15, T. 16 N., R. 1 E.* Northwest side of the buttes, near Keenan's Ranch. Richfield-Buttes Community D-1 Well. Depth, 7,226 feet. Basement described by others as *"gneissic quartz diorite."* Several "sills" of rhyolite were also cut, beginning at a depth of 5,149 feet.

3. *N.W. corner of Sec. 35, T. 16 N., R. 1 E.* The first two gas wells drilled within the Sutter Buttes lie close to the southern edge of the andesitic core, a short distance southwest of South Butte. The initial well, after penetrating Cretaceous sediments, is said to have cut *hornblende-biotite andesite* at a depth of 2,727 feet. Well No. 2 cut "sills" of similar andesite intrusive into Cretaceous shales at various

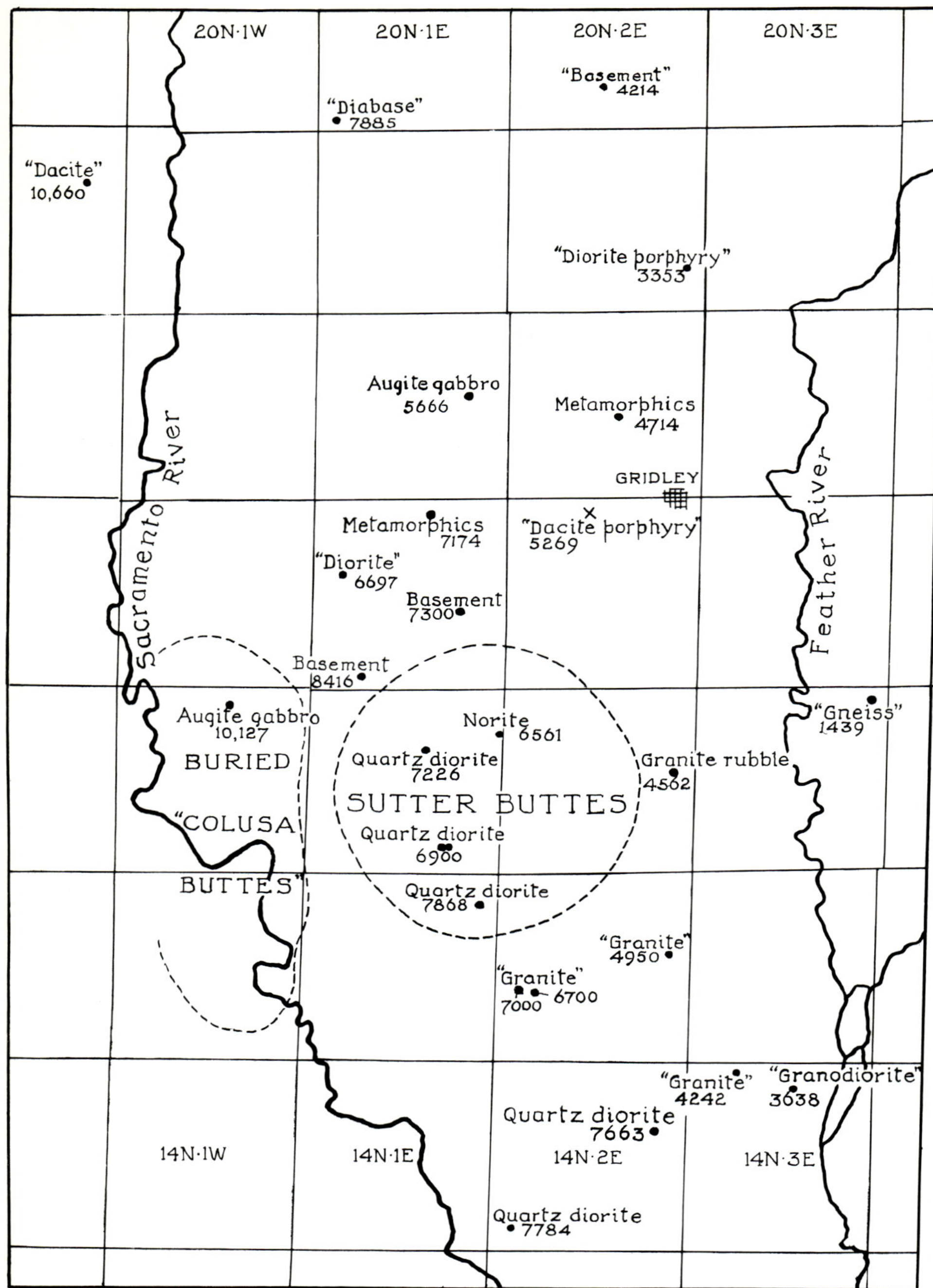

Fig. 3. Depths to the plutonic-metamorphic basement within and near the Sutter Buttes. "Dacite" and "dacite porphyry" are probably of Pliocene or Pleistocene age.

depths between 2,900 and 4,983 feet, and again at intervals between 5,929 and 6,650 feet. At a depth of 6,900 feet, the well cut fine-grained *quartz-free hornblende diorite.* Between 6,917 and 6,927 feet, it cut *maroon cherts and quartz-chlorite schists.* Thence to the bottom, at a depth of 7,014 feet, the well cut medium- and coarse-grained *hornblende- and quartz-hornblende diorites,* their content of quartz ranging from less than 5 to approximately 40 percent. (Core-samples and thin sections in Univ. Calif. Berkeley Coll. 134.) The quartz-hornblende diorite from the bottom of the well yielded a *K/A age of 138 m.y. (Late Jurassic).*

4. *Sec. 12, T. 15 N., R. 1 E.* Close to the southern edge of the buttes. Richfield-Buttes Well No. 14. At a depth of 7,868 feet, the well cut *hornblende-biotite quartz diorite.*

South of the Sutter Buttes

1. *Sec. 19, T. 15 N., R. 2 E.* Approximately 2 miles south of the buttes. Atlantic Epperson Well No. 1. Cut *"weathered granitic basement"* at a depth of 7,000 feet. A second well, about half a mile to the east, is reported to have cut similar material at a depth of 6,700 feet.

2. *Sec. 31, T. 14 N., R. 2 E.* Roughly 10 miles south of the buttes. According to a report in Calif. Div. Mines and Geol. Bull. 181, 1962, this well cut *quartz diorite* at a depth of 7,827 feet; another report places the depth at 7,784 feet.

3. *Secs. 3 and 5, T. 14 N., R. 3 E.* Between 6 and 7 miles southeast of the buttes. Two wells cut *"granite"* and *"granodiorite"* at depths of 4,242 and 3,638 feet, respectively.

4. *Sec. 13, T. 14 N., R. 2 E.* Approximately 9 miles southeast of the buttes. Shell-Sheffield Well No. 1. Cut *hornblende-quartz diorite* at 7,663 feet. (Sample and thin section in Univ. Calif. Berkeley Coll. 134-102.)

5. *Sec. 24, T. 13 N., R. 2 E.* Approximately 15 miles south of the buttes. *"Diorite"* cut at a depth of 7,507 feet.

6. *Sec. 29, T. 12 N., R. 4 E.* Approximately 20 miles south-southeast of the buttes. Sunray-Lienert Well No. 55. Cut *hornblende-biotite quartz diorite* at a depth of 6,033 feet. The hornblende was dated by the K/A methods as *121 m.y. old* (Early Cretaceous). (Thin section in Univ. Calif. Berkeley Coll. 134-100.)

7. *Sec. 10, T. 7 N., R. 3 E.* Approximately 9 miles southwest of the City of Sacramento and 7 miles west of the Sacramento River. G. P. Glide Well. Cut *ophitic, uralitized augite diabase* at a depth of 15,300 feet. (Thin section in Univ. Calif. Berkeley Coll. 134-100.)

Upturned Cretaceous and Tertiary Beds

Introductory Notes

Our main concern is with the volcanic history of the Sutter Buttes; hence, it must suffice to present a brief account of the Cretaceous and Tertiary sedimentary beds, particularly because Williams (1929) has already described them in some detail. It seems desirable, however, to correct several errors in that earlier paper and to include new information gathered by petroleum geologists. Our object is to repair parts of the frame for our new picture of the Buttes' ancient volcanoes.

All the Cretaceous beds were formerly regarded as part of the Chico formation. Discovery of gas fields within the buttes and surrounding Sacramento Valley has since led to detailed subdivision and re-naming of these Late Cretaceous deposits and to clearer understanding of their relations to overlying Tertiary beds. Especially noteworthy from both structural and stratigraphic points of view has been the discovery that the principal bed of white sands in the Sutter Buttes is not of Eocene age, as formerly supposed, but Cretaceous. When Williams mapped the area in 1926-1927, Victor Allen, a fellow graduate-student, was studying the Eocene Ione formation in the foothills of the Sierra Nevada. There seemed no

doubt to either investigator that the main deposit of white quartz-anauxite sands in the Sutter Buttes was the precise equivalent of the fluviatile and estuarine white sands of the Ione formation at its type-locality in the Sierran foothills. Williams and others had found *Trigonias* and other Cretaceous fossils above the main deposit of white sands in the Sutter Buttes, but the mineral-content of the sands was so strikingly similar to that of the Ione sands of the Sierran foothills that the fossils were wrongly assumed to have been reworked. Our own re-mapping, and field work by many petroleum geologists, have amply demonstrated that well above the main deposit of white sands there is a much thinner, discontinuous bed of almost identical appearance, and it is this that corresponds to the Eocene Ione beds of the Sierran foothills. Accordingly, the main deposit of white sands within the buttes, important as one of the principal gas reservoirs, is referred to nowadays as the Kione formation.

We also note in this historical review that Williams erred in his 1929 paper in supposing that certain buff sands and greenish glauconitic and foraminiferal shales of Eocene age, which he called the Marysville formation, underlie the white sands now called Kione. It is now apparent that these beds, currently included in the Capay formation, overlie the white sands of the Kione but underlie those of the Ione formation.

We emphasize in conclusion that widespread landslides, strong lithological similarities between many Cretaceous and Tertiary beds, and the occurrence of numerous faults combine to make mapping of the sedimentary moat extremely difficult, particularly on the east and west flanks of the buttes. Contacts shown on our geological map (fig. 4) should therefore be regarded as only approximate. Almost all the sediments were laid down in shallow, migrating seas or in deltaic flats and lagoons, and some were laid down on broad floodplains. Little wonder, therefore, that beds vary considerably in thickness and lithology, even over short distances, and no doubt some were formed by reworking of older sediments.

Garrison (1962) considered the sedimentary sequence within the buttes to be as follows:

		Thickness in feet
Mio-Pliocene	Sutter formation	c. 1,800
	Unconformity	
	Undifferentiated unit	
Middle Eocene	Ione sand	c. 400
	Capay shale	
Upper Cretaceous	D zone (subsurface)	
E zone*	Kione sand	c. 700
F, F^1 zones*	Forbes shale	2,700-4,000
G 1 zone	Funks shale	c. 1,300

*Zone-names after Goudkoff (1945).

Cretaceous Beds

No Lower Cretaceous beds are exposed and none are cut by wells drilled within the Sutter Buttes. Along the west side of the Sacramento Valley, however, only 20 miles from the

buttes, Lower Cretaceous shales, sands, and conglomerates range in thickness from approximately 16,000 to 20,000 feet. Only the upper part of the Upper Cretaceous sequence is present within the buttes and only to a maximum thickness of about 5,000 feet, but Upper Cretaceous beds in the Coast Ranges are roughly 13,000 to 14,000 feet thick (Lachenbruch, 1962, fig. 2). Moreover, Upper Cretaceous beds continue to thin rapidly to the east of the buttes until they disappear at the edge of the Sierra Nevada.

Some wells within the Sutter Buttes penetrate Funks shale, but the only exposed Cretaceous beds are the overlying Forbes shale and Kione sands.

1. *Forbes Shale.* Included under this category are all the exposed Cretaceous beds below the white Kione sands. They are best seen along the south side of the buttes where most of them stand vertically or almost so. Shales predominate but there are many beds of micaceous and ferruginous sandstone, some of sandy limestone, and a few lenses of conglomerate. The proportion of shale generally increases downward, and in deep wells shales greatly predominate over other beds.

2. *Kione sands.* These rest conformably on Forbes shale and consist typically of white quartz-anauxite sands, though some beds are tinted in shades of pink, purple, and brown by oxidation of small amounts of iron. Intercalated between the sands are lenses that carry abundant pebbles and cobbles of milky vein-quartz, quartzite, and varicolored cherts derived from bedrocks in the Sierra Nevada. In addition, there are lenses of compact ferruginous and calcareous siltstone and sandstone some of which carry *Trigonias* and other Cretaceous fossils.

The thickness of the Kione sands varies markedly over short distances. On the south side of the buttes, their exposed thickness is rarely more than 150 feet, but in wells only a mile or so south of the sedimentary moat, beds grouped by petroleum geologists as Kione are reported to be more than 600 feet thick. Kione sands also appear to be much thicker on the west side of the buttes than on the north. These observations, and the common occurrence of cross-bedding, suggest that many Kione beds are channel-filling deposits laid down in shallow seas and on tidal flats.

Microscopic study of the sands shows that in addition to the predominant grains of quartz, quartzite, and chert and the "silvery" flakes of anauxite, there are occasional grains of hornblende, biotite, and microcline as well as chips of graphitic and micaceous schists and of metabasalt or metadiabase, all of which suggest provenance among plutonic and metamorphic terrains of the Sierra Nevada.

Eocene Beds

A pronounced unconformity separates Eocene from Cretaceous beds. Garrison (1962) divided the former into a lower member (Capay Shale), a middle member (Ione Sand), and an undifferentiated, un-named upper member. Their aggregate thickness seemed to him to be about 400 feet, and their age to be Middle Eocene. However, Lachenbruch (1962) and others assign the Capay beds to the Early Eocene, and our colleague, Wyatt Durham, thinks that some Ione beds at the type-locality may also be of this age.

Beds that Williams (1929) grouped as the Marysville formation, supposedly of Meganos age, are now assigned to the Capay formation, and beds that he called Ione and thought to be of Eocene age are now assigned to the Cretaceous Kione formation. The thin, discontinuous lenses of true Ione (Eocene) sands lie close to the base of the beds that Williams grouped together as the Butte Gravels.

1. *Capay Shales.* These Early Eocene sediments consist chiefly of buff sands locally rich in ferruginous concretions, and glauconitic shales rich in foraminifera. Carbonaceous mudstones are occasionally present and there are thin seams of inferior coal of limited extent on the north and east side of the buttes. The assemblage as a whole suggests deposition in shallow seas and small lagoons. The thickness of the beds on the west side is about 400 feet; elsewhere it varies considerably but nowhere exceeds that amount. By contrast, the thickness of the Capay beds at their type-locality in the Coast Ranges approximates 2,500 feet (Hackel, 1966).

2. *Ione Sands and Butte Gravels.* Williams gave the name "Butte Gravels" to a series of gravels with interbedded sands and thin lenses of limestone and sandstone generally eroded to form round-topped hills and ridges (plate 3a) and yielding thick, clayey soils heavily charged with pebbles and cobbles. Small, widely spaced outcrops of white quartz-anauxite sands, a few feet to a few tens of feet thick, are present beneath the main bed of gravels; these are now considered to be equivalents of the white sands at the type-locality of the Ione formation in the Sierran foothills.

The Butte Gravels (*sensu lato*) are shallow marine deposits with subordinate fluviatile and deltaic interbeds, and hence vary both in thickness and lithology over short distances. They reach their maximum thickness, about 1,200 feet, on the south side of the buttes, but on the west they are only about half as thick, and elsewhere are rarely more than 400 feet thick. Particularly variable are the gravel- and boulder-lenses. On the northwest side of the buttes, near the old Keenan's Ranch house, cobbles and boulders up to a foot across are abundant, but on the west, cobbles and boulders are only present locally. Indeed, on this side of the buttes, it is often difficult to distinguish finer beds of the "Butte Gravels" from some Cretaceous sediments and some pebbly layers in the Sutter formation. Coarse gravels are again widespread on the south and southeast sides of the buttes. All occupy irregular channels within sandy and silty beds.

The nature of the pebbles, cobbles, and boulders leaves no doubt that all were derived from bedrocks in the Sierra Nevada. In order of abundance, the rock-types are as follows: quartz porphyries and metarhyolites, milky vein-quartz, varicolored cherts, quartzites, dense greenstones, granodiorites and quartz diorites, and various kinds of siliceous schists.

On the south side of the buttes, the pebbly and cobbly beds are separated by ferruginous sands and clays, and locally by two conspicuous, hard beds, each approximately 4 feet thick, one a gray limestone, roughly 150 feet above the top of the Kione sands, and the other a brownish, richly fossiliferous sandstone, 700 feet above the Kione. The rich fauna from this sandstone, which underlies the main bed of gravel, is dominated by *Turritella merriami,* Dickerson. Bruce Clark considered it to be of Middle Eocene age, but it is now regarded as Early Eocene.

The overlying coarse Butte Gravels are marine equivalents of the fluviatile "Auriferous Gravels" of the Sierra Nevada. Their age within the Eocene is uncertain. Harry McGinitie (personal communication) considers the flora from the upper part of the Auriferous Gravels to be Early Eocene. Dalrymple (1964) says that "a middle Eocene age for some of the Sierran auriferous gravels in or near the foothills seems to be well established; however, many of the seemingly prevolcanic auriferous gravels are Oligocene and perhaps some are as young as early Miocene." For the present, therefore, it seems best to say that the Sierran auriferous gravels range in age from Middle Eocene to and including Oligocene; however, some intra-volcanic "Bastard Gravels" are certainly of Miocene age. As for the Butte Gravels in the Sutter Buttes, we can say unequivocally that they are pre-volcanic and that all are probably of Eocene age.

Oligocene, Miocene, and Pliocene (Sutter) Beds

The Sutter formation generally lies conformably, but locally with marked unconformity, on underlying Tertiary beds. It consists almost exclusively of volcanic sediments carried down by rivers from the Sierra Nevada to be deposited in deltaic fans and on broad floodplains that occupied most of the Sacramento Valley during Oligocene, Miocene, and Pliocene times. In brief, the formation is made up chiefly of outwash from the voluminous andesitic lahars (volcanic mudflow deposits) that once blanketed virtually all the western slopes of the Sierra Nevada north of Yosemite. Mingled with the volcanic debris are materials derived from the "Auriferous Gravels" that underlie the Sierran lahars; and, in a few places near the base of the Sutter formation, there are thin lenses of white, airborne rhyolitic, pumiceous ash, equivalents of some of the Oligocene-Miocene (Valley Springs) rhyolitic deposits on the flanks of the Sierra directly east of the Sutter Buttes.

The name "Sutter formation" was first applied to these volcaniclastic sediments by Dickerson (1913). Since then, however, most petroleum geologists have substituted the name "Tehama formation" in the mistaken belief that the Sutter beds are identical with the volcaniclastic, continental sediments along the west side of the Sacramento Valley to which Russell and Vanderhoof (1931) gave the name "Tehama." In our opinion, this practice is unfortunate because the Sutter beds are only in part of Pliocene age, whereas all or almost all of the Tehama beds are said to be of this age. Moreover, as we have already noted, the Sutter beds were derived from volcanic deposits in the Sierra Nevada whereas almost all the debris in the Tehama formation was derived from the Coast Ranges to the west and Cascade volcanoes to the north.

Harry Johnson (1943) called the pre-volcanic conglomerates near the top of the Butte Gravels the basal member of the Sutter formation, but this was a mistake. The basal member on the south side of the buttes is represented by a few small outcrops of white rhyolite tuff. For instance, a short distance east of the De Witt volcano (Sec. 6, T. 15 N., R. 2 E.), there are thin lenses of rhyolitic pumice made up of such delicate, fibrous fragments that they must have fallen into place directly from the air. They must have been blown from volcanoes in the Sierra Nevada and be equivalents of some of the rhyolite tuffs of the Valley Springs formation, widespread on the Sierran slopes beneath the andesitic lahars of the Mehrten formation.

To test this hypothesis, we examined exposures of Valley Springs beds on the Sierran slopes east of the Sutter Buttes. Well-bedded fluviatile deposits of rhyolite tuff outcrop beneath andesitic lahars at an elevation of about 3,000 feet, roughly 2 miles east of Nevada City, along the old road (No. 20) to Emigrant Gap. Roughly 13 miles farther east along the same road, at an elevation of approximately 4,500 feet, fluviatile rhyolite tuffs are capped by more compact, well-jointed tuffs rich in crystals of quartz, feldspar, and biotite, and containing many flattened, flamelike lapilli (*fiamme*) of pumice. These compact tuffs are either deposits of hot lahars or, more likely, products of glowing avalanches. Elsewhere on the neighboring slopes of the Sierras, firmly welded ash-flow tuffs (ignimbrites) are interbedded with fluviatile deposits of similar composition. There can be no doubt, therefore, that from time to time during accumulation of the Valley Springs beds, torrents of pumiceous ash swept down the Sierran slopes to spread across broad floodplains that then occupied most of the Sacramento Valley, and showers of airborne rhyolitic ash, erupted prior to many glowing avalanches, fell as far away as the Sutter Buttes.

As to the age of the basal rhyolite tuffs of the Sutter formation, we can only say that it lies between about 20 and 30 million years, this being the approximate age-range of the Valley Springs rhyolite tuffs, as determined by Dalrymple (1964); in other words, their age is Early Miocene or Late Oligocene.

Above the lenses of pure, airborne rhyolite tuff at the base of the Sutter formation lie waterlaid beds that consist of a mixture of biotite rhyolite and pyroxene and andesite debris. But by far the bulk of the Sutter formation consists of waterlaid andesitic material. A few beds may have been laid down in shallow seas, but the widespread occurrence of cross-bedding and the apparent absence of fossils, other than horse-teeth, incline us to the view that almost all of the Sutter beds were laid down by streams.

Lithological and mineralogical characters of the sediments have been described already (Williams, 1929). It is enough to repeat that all but a few Sutter beds represent fine outwash from the copious "bouldery volcanic mudflows" (lahars) of the Mehrten formation on the slopes of the Sierra Nevada directly east of the Sutter Buttes. Admixed with the andesitic detritus is abundant pebbly "bedrock debris," chiefly of quartz, quartzite, chert, schist and plutonic rocks carried down to the Sacramento Valley by Sierran rivers, principally by the Tertiary Yuba River, during the long volcanic cycle. A similar mixture of volcanic and non-volcanic debris may be seen among the "Bastard Gravels" near the edge of the Sierra Nevada around Smartsville and Sicard Flat, 20 to 25 miles east of the Sutter Buttes. In brief, most of the Sutter formation consists of andesitic and rhyolitic debris mingled with pre-volcanic materials derived by reworking of the "Auriferous Gravels" of the Sierra Nevada.

In view of how the Sutter beds originated and were deposited, it is not surprising that they vary greatly in thickness. Harry Johnson (1943) thought that locally they are more than 1,500 feet thick, but it should be recalled that both Johnson and Garrison included in the Sutter formation some pre-volcanic pebbly beds that belong to Williams' Butte Gravels. The thickness of the Sutter beds, in most places is much less than they supposed. Along the south side of the buttes, it approximates 1,000 feet; along the west side, it approximates 600 feet.

The age of all but the basal rhyolitic beds is Mio-Pliocene; virtually all are coeval with the andesitic mudflow deposits (lahars) of the Sierra Nevada, and hence range in age from about 10 to 20 million years. However, deposition of reworked Mehrten beds continued in the Sutter Buttes until the onset of local volcanism, approximately 2.0 to 2.5 million years ago. Fragmentary teeth found about 250 feet from the top of the Sutter beds south of the West Butte Pass road (N.E. corner of Sec. 3, T. 15 N., R. 1 E.) were identified by our colleague, Professor Donald Savage, as *Dinohippus mexicanus*, a form very closely akin to the modern horse, and assigned by him to Latest Pliocene (Late Hemphillian) time, approximately 4 to 5 million years ago.

Pre-Explosive Uplift of the Sedimentary Beds

Introductory Statement

Extrusion of the dacite dome of Syowa Sinzan in Hokkaido in 1944-1945 was preceded by uplift of flat-lying beds to form a "Roof Mountain" approximately 650 feet high and

1,100 yards across (Minakami *et al.*, 1951). Uplift took place within about a year, accompanied during late stages by explosive eruptions that blasted many small craters through the uplifted beds before viscous magma emerged to build a summit-dome of Pelean type. A somewhat similar sequence of events took place repeatedly at the Sutter Buttes, but on a vastly greater scale and over a vastly longer span of time. Rising magmas uplifted Cretaceous and Tertiary beds within a circular area more than 8 miles across. As uparching progressed, the beds were broken by roughly radial and concentric faults and were subjected to rapid erosion. Not until almost all the Tertiary beds had been stripped from the central part of the buttes did explosive activity begin. Explosions then recurred at short intervals for more than half a million years, accompanied by the rise of many domes of viscous magma, some of which broke through to the surface while others solidified underground. Deformation of sedimentary beds continued throughout the explosive episode, but our main concern in this section of the report is with the regional, pre-explosive uplift.

Williams maintained in his 1929 paper that regional uplift was produced by a central, andesitic laccolith prior to extrusion of the rhyolite domes within the surrounding sedimentary moat, and he estimated that the laccolith uparched its cover to form a mountain 7,000 feet high. These opinions call for drastic revision. Many gas wells, numerous geophysical measurements, and closer study of structures within the Rampart formation have provided a wealth of new information. It is now clear that regional uplift was not caused by a simple laccolith of andesite, but by a group of intrusions, probably composed mostly of rhyolite like those that domed deeply buried Cretaceous sediments in the adjacent "Colusa Buttes." And these rhyolitic intrusions may have been followed by others of dacitic and andesitic composition. Early rise of light, but extremely viscous rhyolitic magma may well have caused most of the initial, pre-explosive, regional uplift in the Sutter Buttes. It must be emphasized, however, that until more radiometric measurements have been made on both exposed and buried intrusive bodies, it remains difficult or impossible to separate uplifts caused by pre-explosive intrusions from those developed during the explosive phase. A broad, regional uplift already existed before explosions began, but more localized and generally more pronounced uplifts took place during the explosive phase when the sedimentary beds upturned in the moat had been eroded almost to their present levels.

There was never a circular mountain 7,000 feet high prior to the first explosions, as Williams once supposed. All but a few of the andesitic peaks that now rise in the interior of the buttes are Pelean domes that grew near the close of explosive activity. In the Henry and La Sal Mountains of Utah, emplacement of central stocks was accompanied or followed by intrusion of peripheral laccoliths, but in the Sutter Buttes central intrusions were emplaced long before most of the surrounding, satellitic domes were emplaced.

When regional uparching began and how long it lasted, cannot be told; but the oldest of the exposed fragmental ejecta at the base of the peripheral Rampart beds may be less than 2.5 million years old. Available well-data tell us nothing about the volume of sedimentary debris removed by erosion during the slow uplifts before the first explosions. Such debris, consisting mostly of reworked Sutter beds, must have been penetrated by many gas wells, but no records have come to our attention. Regional uplift was attended by extensive landslides and rapid erosion during pluvial episodes of Late Pliocene and Early Pleistocene times.

As for the shapes and sizes of the intrusive bodies now concealed beneath Pelean domes in the center of the buttes, we can do no more than speculate, as we have ventured to do in our cross-sections (figs. 6, 7, and 8). Some intrusive bodies were probably plug-like; others were sills or transgressive sheets like those cut by wells in the buried Colusa Buttes. Some intrusions may have been bordered by smaller, crudely radial, inward-dipping offshoots. The propulsive forces that caused the initial, pre-explosive uparching depended on the depth of origin and volume of the ascending magmas and on their density and viscosity. That the magmas were extremely viscous is obvious from the extensive deformation they produced.

It seems appropriate to digress briefly to indicate some similarities and differences between these initial intrusions and the better known ones in the Henry and La Sal Mountains of Utah, so well described by Gilbert (1877), Hunt (1953), Hunt and Waters (1958), and Johnson and Pollard (1973). The central stocks in those mountains consist mostly of diorite- and monzonite-porphyries and are surrounded by radiating sills most of which swell at their tips to form laccoliths. Structural uplifts around the central stocks are much more symmetrical than at the Sutter Buttes, and generally range between 3,500 and 7,000 feet in height. Gilbert erred when he envisioned the central stocks in the Henry Mountains having risen, piston-like, between vertical circular faults; later studies have shown that there are no such bounding faults around either the Henry Mountains or the La Sal Mountains stocks, nor are there any around the central intrusions of the Sutter Buttes. But in all three places, the intrusions were too cool to cause any metamorphism beyond a slight baking of adjacent shales.

In the Henry Mountains, most central stocks are surrounded by shatter zones, up to a mile in width, in which igneous and indurated sedimentary fragments are intimately admixed and tilted at all angles, and locally are so intensely crushed as to produce veinlike masses of pseudo-tachylyte. There is no such halo of shattered rocks around any intrusions in the Sutter Buttes; on the contrary, all igneous-sedimentary contacts are clean-cut.

None of the laccoliths in the Henry Mountains broke through to the surface, and probably none of the central stocks fed explosive eruptions. In the La Sal Mountains, however, one central stock developed pipelike explosion-vents during closing stages of activity. In the Sutter Buttes, on the contrary, the initial intrusions were followed by long-continued volcanism, involving the protrusion of many Pelean domes and voluminous explosive eruptions.

Faulting Accompanying Uplift

Although the central stocks of the Henry and La Sal Mountains caused more extensive upturning of adjacent sedimentary beds than did the central intrusions of the Sutter Buttes, they appear to have caused little or no faulting. In the Sutter Buttes, on the contrary, the pre-explosive intrusions not only uparched adjacent beds but broke them into blocks bounded by roughly radial and concentric faults (fig. 4 and structure-contour map, fig. 11). Additional faults, not exactly located, underlie alluvium in the large radial valleys that cut both the sedimentary moat and surrounding rampart. Many faults developed long after the initial, regional uplift; indeed some did not develop until near the close of the explosive episode. Moreover, some faults, including the north-south Brockman Canyon fault on the south side of the buttes, formed during the initial arching and were reactivated during the explosive episode so as to displace lower and middle members of the Rampart formation.

We are much indebted to the Buttes Gas and Oil Company for the valuable structure-contour map prepared by E. D. Sherman (fig. 11). This reveals the presence of several domical uplifts beneath the smooth slopes of the rampart, caused by buried intrusions, chiefly of rhyolite; it also indicates several faults concealed beneath the rampart, some of which must be assigned to the pre-explosive episode, while others, including those that underlie the De Witt volcano, formed during the explosive episode.

Height of the Pre-explosive Uplift

How high were the Sutter Buttes before explosive volcanism began? No accurate estimate is possible, but certainly Williams (1929) exaggerated greatly when he originally envisaged a domical mountain 10 miles wide and 7,000 feet high, similar in shape and origin to the well-known laccolithic uplift of Navajo Mountain in northern Arizona.

The visible margins of the igneous core of the buttes range in elevation from 300 to 1,300 feet, but all the high andesitic peaks within the core (e.g. North, South, and West Buttes, and Twin Peaks) are Pelean domes that rose long after the pre-explosive uplift. It seems probable, indeed, that none of the pre-explosive intrusions rose to sea-level. Moreover, they rose to different heights so as to uparch the sedimentary beds differentially, and hence may have produced a large east-west depression on the present site of Bragg's Canyon, where a lake existed during part of the ensuing explosive episode. Bragg's Canyon, in other words, may owe its origin to easy removal of sedimentary and volcanic debris from an initial, throughgoing depression.

The sedimentary cover above the initial intrusions was much thinner than Williams formerly supposed. No inliers of Cretaceous or Tertiary beds remain in the central part of the buttes, and if the beds upturned in the encircling moat were depressed to their original horizontal position they would occupy only a small fraction of the area now occupied by the igneous core. Accordingly, a vast amount of the sedimentary cover must have been removed before explosive activity began. Steeply dipping Cretaceous and Tertiary beds, between 4,500 and 5,000 feet thick, adjoin the igneous core on the south, but not all of these arched over the core. Uplift was a long-continued process, accompanied by rapid erosion. Moreover, the steep dips were produced mostly by intra-explosive uplifts. South of West Butte Pass, for example, the basal Rampart beds and underlying Sutter beds are locally conformable and, though initially laid down horizontally, now dip southward at angles of 10° to 30°. An andesitic intrusion cut by the first gas well, close to the edge of the igneous core of the buttes, south of South Butte, was emplaced 1.74 m.y. ago, i.e. during the explosive episode. Andesite cut by the adjacent, second gas well is probably of the same age, and one cut by a well approximately a mile southwest of South Butte may also be of this age. Accordingly, the steep dips of the sediments south of South Butte are probably due chiefly to intrusions emplaced long after the initial, pre-explosive uplifts (see notes on "Intra-volcanic Deformation," below).

Close to the east and west sides of the core, exposed Cretaceous beds dip outward generally at angles of about 10° to 30°; some of these and overlying Tertiary beds must once have covered the igneous core, but to what depth cannot be judged without knowing how much uplift was caused by intra-explosive intrusions and how fast uplift was in relation to the rate of erosion. Along the north side of the buttes, most of the exposed sediments are Tertiary and outcrop at elevations of no more than 600 feet. Except where deformed by later, intra-explosive uplifts, they generally dip outward at about 10°. Hence, the northern part of the core can never have been covered by sediments to a depth of more than approximately 1,200 feet, and if erosion was rapid during uplift, the cover may have been considerably thinner or preserved only in patches.

Our conclusion is that before explosive activity began, the Sutter Buttes were highest on the south, east, and west sides where they may have reached 2,500 feet. They were considerably lower on the north side, and were lowest on the present site of Bragg's Canyon where a large lake developed during the succeeding explosive phase.

As to how fast the sedimentary cover was stripped from the rising core before the initial explosions, we can only speculate. Our inclination is to think that the interval between the regional uplift and first explosions was relatively brief, perhaps no more than a few tens of thousands of years. Few volcanoes lie dormant for such protracted periods.

Some Detailed Evidence

We now present supporting evidence concerning the pre-explosive uplift, beginning on the south side of the buttes and continuing in a clockwise direction; in doing so, we cannot avoid discussing many of the intra-explosive uplifts.

South side. In the sedimentary moat south of South Butte, Cretaceous and Tertiary beds, originally horizontal and with an aggregate thickness of almost 5,000 feet, are upturned to verticality and even overturned by as much as 20°. The prevailing strike of the beds within this steeply dipping sector is east-west, though strikes swerve sharply in several places because of buried intrusions. One such intrusion was cut at a depth of approximately a mile by a gas well a mile southwest of South Butte (S.W. ¼ Sec. 35, T. 16 N., R. 1 E.). It consists of hornblende-biotite andesite similar to most of the andesites in the core of the buttes. Other buried intrusions within the sedimentary moat are composed of biotite rhyolites similar to some of the intra-explosive bodies exposed at the surface.

The steep dips within the moat south of South Butte were certainly caused mainly by intermittent uplift during the explosive phase. Elsewhere, we noted that south of the West Butte Pass road, the basal Rampart beds are locally conformable on Sutter beds, both dipping southward at angles of 10° to 30°; we note also that north of the road, approximately a mile south-southwest of South Butte, there is a small outlier of the basal member of the Rampart formation in which the beds dip southward at about 30°, a sure sign of continued, intra-explosive uplift. (See fig. 5.)

West side. The Cretaceous beds adjoining the igneous core on this side of the buttes generally dip outward at angles of about 20°, but farther from the core both the Cretaceous and Tertiary beds are upturned to much higher angles, commonly to about 70° and locally to verticality. The zone in which the dips suddenly steepen coincides with a belt of concentric faulting. A gas well (Richfield-Buttes Community C well) drilled in this belt (near the center of the north half of Sec. 28, T. 16 N., R. 1 E.) cut andesite at a depth of 6,133 feet. The concentric faulting and sudden steepening of dips may therefore be related to a large, buried intrusion of andesite, but whether or not this was emplaced before explosive activity began is uncertain.

A mile and a half west of West Butte, on "547 Hill," Sutter beds dip westward at 30° to 40° though overlying Rampart beds, which rest on a deeply channeled surface, dip in the same direction at angles of only 5° to 10°. Here, as on the south side of the buttes, is clear proof of intra-explosive deformation following the initial, regional uplift. Moreover, a concentric fault near the west edge of Sec. 28, T. 16 N., R. 1 E. passes close to the base of "547 Hill" and swings northeast along the head of the large alluvial valley east of Noyesburg School (see structure-contour map, fig. 11). Locally, Sutter beds west of the fault stand vertically or almost so.

A short distance to the south, in Fig Tree Gulch (near the boundary between Secs. 28 and 33), another concentric fault separates gently dipping Cretaceous beds on the east from much more steeply dipping Tertiary beds on the west. Much more work is necessary before the complicated structures on this side of the buttes are unravelled.

Northwest side. Near the old Keenan ranch-house (N. ½, Sec. 15, T. 16 N., R. 1 E.), Tertiary sediments generally dip west and northwest at angles of approximately 20°, whereas the Rampart beds dip at angles of 10° or less in the same direction. A short distance to the

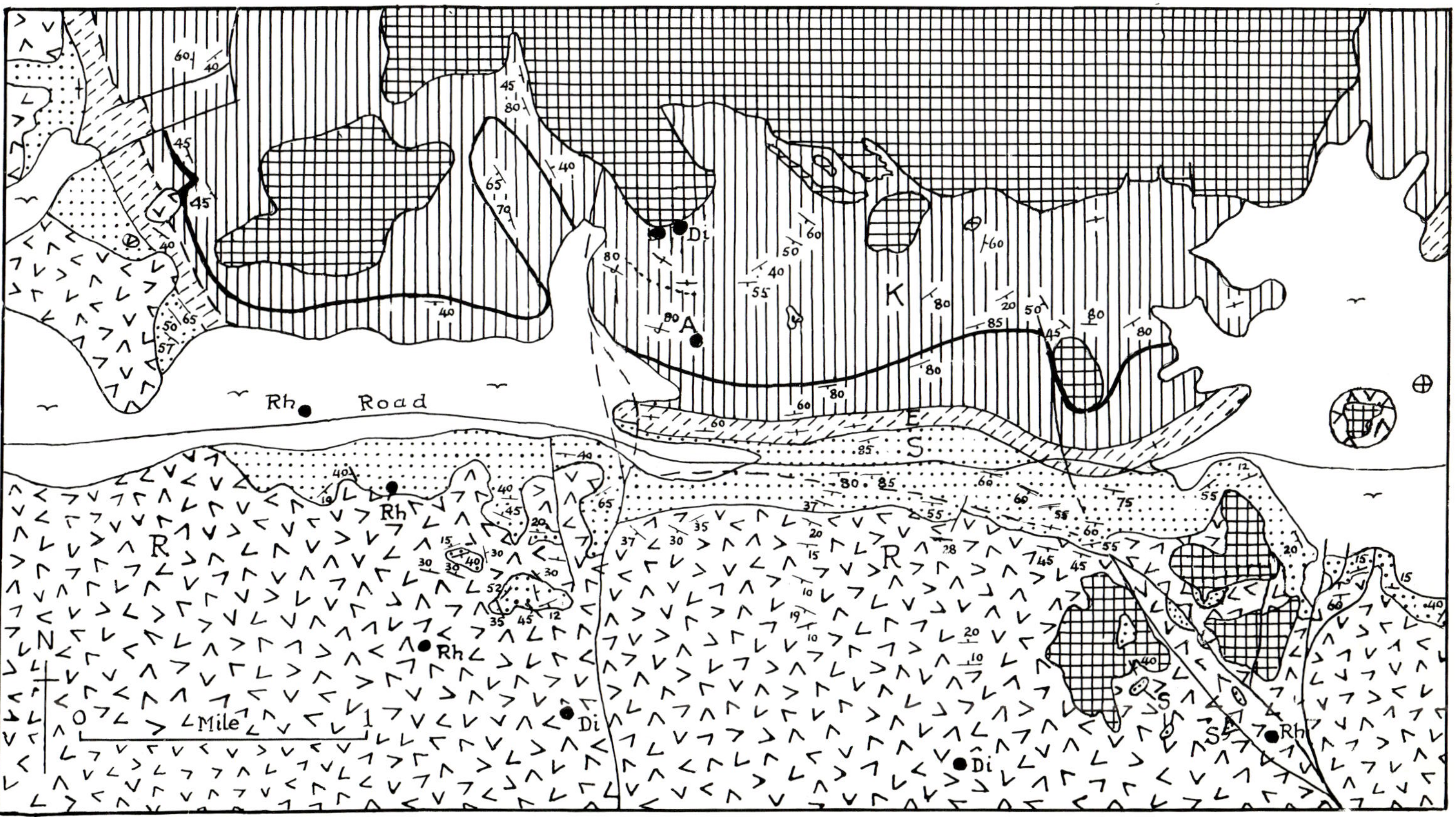

Fig. 5. Geologic map of the area adjoining West Butte Pass road, on the south side of the buttes. Cross-pattern denotes andesite and rhyolite domes and flows; chevron-pattern denotes Rampart formation. "K" stands for Cretaceous beds, including the Kione bed, indicated by heavy black line; "E" for Eocene beds; "S" for Sutter beds. Rocks cut by drilling, as follows — "A" for andesite; "Rh" for rhyolite; "Di" for diorite.

north (in Sec. 10), near the nose of the plunging anticline cut by the Richfield-Buttes Community D-1 well, the angular unconformity between the Sutter and Rampart beds increases to about 30°. Pre-explosive uplift must have continued hereabouts during the explosive phase. Half a mile east of the Richfield-Buttes well, the basal Rampart beds dip north and northwest at 40° to 50° whereas the overlying, channel-filling Rampart beds dip only 10° or so in the same direction.

North side. In Peace Valley, Tertiary beds were tilted outward at angles of only about 10° during the initial, pre-explosive uplift; and the topmost Sutter beds had already been deeply eroded before they were buried by the earliest volcanic debris of the Rampart formation. At the east end of the valley, the Sutter-Rampart contact lies at elevations of approximately 600 feet, but within less than half a mile to the west, where the road from Pennington enters the sedimentary moat, the contact drops to 300 feet. Various units of the Rampart sequence thus overlap on to a deeply channeled surface of Sutter beds. And whereas the base of the Rampart sequence lies at elevations of only about 250 feet along the northwest side of Peace Valley, it lies at 700 feet on the opposite side, half a mile away.

A conspicuous dome of rhyolite rises from the floor of the valley, close to Pugh Place. Half a mile to the east, some Sutter beds are upturned to verticality though adjacent Rampart beds dip only about 5° to the north. Presumably there was strong pre-explosive uplift in this vicinity, possibly caused by a buried extension of the Pugh Place rhyolite dome.

East side. Signs of pre-explosive uparching and erosion are especially well displayed close to the head of Kinch Canyon. Where the canyon leaves the sedimentary moat to enter the encircling rampart (between Secs. 28 and 29. T. 16 N., R. 2 E.), the contact between the Rampart and Sutter beds lies at elevations of approximately 400 feet; yet within half a mile to the northwest, the contact rises to 800 feet; and within the same distance to the southwest, it rises to 700 feet. The Sutter beds must therefore have been strongly uplifted and deeply channeled before explosive activity began.

On the north side of Kinch Canyon, the basal Rampart beds dip north at about 15° though the underlying Sutter beds dip in the same direction at 50° to 60°. Obviously there was pronounced uplift hereabouts both before and during the explosive episode.

Extrusive and Intrusive Domes

The only lava flows within the Sutter Buttes are small ones near the central lake-beds and those that spread over the Rampart beds near Sutter City, during the final phases of volcanism. And the only dike is one of columnar andesite, approximately 3 feet wide, close to the center of the buttes (Sec. 24, T. 16 N., R. 1 E.). With these exceptions, all the rising magmas were so viscous that they were extruded slowly at the surface to build Pelean domes or were intruded at depth forming domical bodies that strongly uparched the surrounding sediments.

Extrusive and intrusive domes range in composition from andesite through dacite to rhyolite, but there seems to be no regular sequence among them. It seems best, therefore, to discuss them in the following order; first the andesitic domes in the center of the buttes and in the surrounding moat; then the intrusive dacites and rhyolites within the andesitic core and the extrusive ones within the moat; and finally dacite and rhyolite domes concealed beneath the moat and encircling rampart.

Andesite Domes

The central part of the buttes is occupied chiefly by andesites, all but a few of which were considered in Williams' 1929 paper to be parts of a single laccolith. We now know, however, that they represent a coalescing group of intrusive and extrusive domes that rose intermittently over a long period. Some andesites were probably discharged before and during early stages of the explosive episode, but all the conspicuous andesite domes rose later, during and after the explosive phase, so as to give rise to the blocky lahars that form the topmost Rampart beds. Two small patches of waterlaid rhyolitic pyroclastic material in the interior of the buttes are the only visible remnants of the numerous ash-cones that were formerly present, though other remnants must lie concealed beneath younger andesitic lavas.

Most of the early andesites have been altered to propylites, whereas the younger ones have been propylitized only in a very few places. Propylites are most widespread south of Peace Valley, particularly in the southern part of Sec. 13, T. 16 N., R. 1 E. and the adjacent part of Sec. 14. They are drab-green rocks forming round-topped, rubble-covered hills with only occasional, small, structureless outcrops. Their topographic forms contrast strongly with the steep-sided ridges and craggy pinnacles typical of the younger, fresher hornblende-biotite andesites. Petrographic notes on the propylites are included in the Appendix; here we note that unlike most propylites in other volcanic regions, few Sutter propylites carry pyrite or any other sulfides.

The interior of the buttes is occupied chiefly by a cluster of coalescing andesite domes of Pelean type, the youngest of which have been modified only slightly by erosion. Most domes are approximately circular, but a few, such as West Butte, are markedly elongate.

North Butte, the youngest and best exposed Pelean dome of andesite, measures roughly 0.7 by 0.5 miles across at the base, and rises to a height of 1,883 feet, approximately 1,000 feet above the adjacent Rampart beds (plate 2a-b; plate 4b). Its steep flanks and summit bristle with vertical slabs and spires. Flow-banding close to and concentric with the margins dips inward, generally at angles of 50° to 70°; nearer the center, most banding is vertical and strikes approximately north-south. Texturally and mineralogically, the andesite resembles that of almost all the other Pelean domes in the interior of the buttes, being characterized by many large phenocrysts of hornblende, biotite, and plagioclase in a dense, usually non-vesicular matrix. And here, as on many other domes, massive andesite is cut by a few irregular, steeply dipping bands and stringers of auto-brecciated material. The lava was extremely viscous as it rose to the surface; close to the eastern base of North Butte some adjacent Rampart beds have been upturned to verticality and some have been slightly overturned.

Another Pelean dome that exhibits its internal structure especially well is the one forming Twin Peaks, about a mile north of South Butte (plate 2a). Flow-planes here are roughly funnel-shaped and generally dip inward at increasing angles toward the middle. Presumably this dome, like all the others, is underlain by a relatively slender feeding pipe.

Another youthful dome that readily attracts attention is the one on the north side of Bragg's Canyon (S.W. ¼, Sec. 14, T. 16 N., R. 1 E.). Throughout most of this protrusion, the flow-planes trend between northwest and west-northwest, standing vertically or almost so, and weathering to produce countless pinnacles and platy stacks.

Only in the South Butte dome did we see horizontal and gently dipping flow-planes, and even here they are restricted to the summit-region.

Isolated bodies of andesite project through the moat of upturned sedimentary beds on both the south and west sides of the buttes. Whether these are remnants of surficial Pelean domes or intrusive masses exhumed by erosion is uncertain. It is clear, however, that many bodies of andesite never broke through to the surface. One such body was cut at a depth of about a mile by a well southwest of South Butte (S.W. ¼, Sec. 35, T. 16 N., R. 1 E; Buttes Gas and Oil Well No. 12); it is reflected at the surface by a sharp swing in the strike of adjacent Cretaceous beds. Andesite was also cut at the bottom of the first gas well drilled in the buttes, close to the margin of the andesitic core south of South Butte, at a depth of 2,727 feet, and again in the second well at various depths down to 6,650 feet. Andesite from the first well is 1.7 million years old; almost certainly, the andesite from the second well is part of the same intrusive mass, and both may be offshoots of the feeder to the extrusive dome of South Butte.

In the moat on the west side of the buttes, andesite was cut at a depth of 6,133 feet. Another well, drilled close to the north edge of the buttes, west of Pennington (Sec. 35, T. 17 N., R. 1 E.) is said to have cut "andesite and rhyolite with phenocrysts of hornblende." We have not seen samples from this well, but it seems unlikely that both kinds of lava are present, and if all contain hornblende they must be either dacites or andesites, not rhyolites.

Only in one place did we observe andesite in contact with rhyolite; this was roughly a mile west of Moore's Ranch House, close to the south edge of the andesitic core of the buttes, and here rhyolite cuts andesite.

All unaltered andesites are remarkably similar both in texture and mineral-content. All contain conspicuous phenocrysts of hornblende, biotite, and plagioclase, and all have a dense pilotaxitic matrix. Petrographic notes have already been published (Williams, 1929) and others are included in the Appendix. Particularly noteworthy is the fact that andesites cut at depths down to more than 6,000 feet are essentially identical with andesites erupted at the surface. Later on, we refer to biotite rhyolites intruded at depths of approximately 9,000 feet in the buried "Colusa Buttes" which are likewise identical in composition and texture to many rhyolites that issued at the surface to form Pelean domes within the Sutter Buttes.

A few andesites in the northern part of the buttes, adjoining Peace Valley, contain occasional basic inclusions up to 6 inches across. These consist of hornblende, diopside, biotite, and calcic plagioclase, and their texture may range, even in a single hand-specimen, from aphanitic to coarsely gabbroid and occasionally to gneissoid. We regard these inclusions as autoliths (i.e. early crystal-segregations in the andesitic magmas) rather than xenoliths torn from the plutonic basement. None have been observed among the youngest andesite domes (e.g. those of North, South, and West Buttes), but they are not uncommon as fragmental ejecta among the basal beds of the Rampart formation.

It should be stressed in conclusion that none of the andesites produced the slightest contact metamorphism of adjacent rhyolites and sedimentary beds beyond a weak induration of some adjacent shales within a fraction of an inch of the contacts. Manifestly, they were emplaced at low temperatures; indeed most of them must have been nearly solid during final stages of their upward movement. Noteworthy also is the extreme scarcity of sulfides.

Exposed Domes of Dacite and Rhyolite

Many irregular siliceous intrusions are present within the andesitic core of the buttes, particularly near the margin, close to South Butte. Other siliceous intrusions must lie concealed beneath the central part of the buttes because dacitic and rhyolitic ejecta are present among the basal deposits of the Rampart formation. Our immediate concern is with the siliceous domes extruded within and along the margins of the sedimentary moat, mostly if not entirely during the explosive episode.

Siliceous domes within the moat are easy to distinguish from afar, not only by reason of their conspicuous topographic forms but also because their whitish colors contrast markedly with the dark brown colors of the weathered andesites (plate 5a). Most of the domes are roughly oval in outline, the major axis of the largest measuring slightly more than a mile in length. It seems certain, however, that many siliceous domes that issued at the surface are only the exposed parts of much larger bodies extending over large areas beneath the surface. All were emplaced in such a viscous condition that they strongly deformed adjacent beds. Explosions of gas rich magma undoubtedly preceded the growth of some domes. For example, the presence of pumiceous biotite rhyolite and rhyodacite tuff, lapilli, and bombs near the base of the Rampart formation close to the northwest base of North Butte is almost surely related to explosions preceding the rise of nearby domes. And though no domes within the sedimentary moat can be seen to rise within explosion-craters, such vents may be buried beneath the domes themselves.

Flow-banding in the outer parts of most siliceous domes is concentric with the margins, and though usually smooth or gently undulating, it may be markedly contorted within the interior of the domes. A funnel-shaped arrangement of flow-planes is not uncommon, but locally the marginal banding may dip steeply outward.

Auto-brecciation is much more widespread among rhyolite domes than it is among dacites and andesites. It is concentrated normally in the marginal parts, but in the elongate dome near De Witt's Ranch house (Sec. 32, T. 16 N., R. 2 E.) brecciation affects almost the entire width of the dome. Some brecciation resulted from differential rates of rise of extremely viscous, partly solidified magma, and much from fracturing of the solidified crusts of domes by continued upsurge of still viscous material within. It seems likely, however, that most brecciation was caused by steam blasts (phreatic explosions) which shattered the rhyolites as they rose through water-soaked sediments. So-called *peperites* formed in this manner have already been described (Williams, 1929, pp. 166-172); they consist characteristically of angular blocks and chips of dense rhyolite in a comminuted, sand-like matrix.

Some peperites were injected as thin dikes into wet sediments bordering the rising domes. Graphic examples may be seen west of the large rhyolite dome near the southwest edge of the buttes (near the center of the boundary between Secs. 28 and 32, T. 16 N., R. 1 E.). Anastomosing, irregular dikelets of peperite, from a fraction of an inch to an inch or two across, cut white Kione sands that had previously been broken into closely spaced blocks by faults with displacements of a few inches to a few feet. The peperites consist of angular chips of rhyolite, from sand-size to about 6 inches across, mingled with bits of shale and occasional pebbles. They must have been injected into wet Kione beds as these were being uparched and broken by rising rhyolitic magma.

Petrographically, the rhyolites are typified by numerous phenocrysts of biotite, quartz, and plagioclase, and by subordinate ones of sanidine. Most dacites are also strongly porphyritic, but they lack sanidine and generally contain phenocrysts of hornblende with or without flakes of biotite. But both rhyolites and dacites have dense, micro- to crypto-felsitic matrices. Indeed, the margins of a few domes may have been so rapidly chilled by contact with water-soaked sediments that they were originally glassy. Black, devitrified obsidian may be seen, for example, at the southern end of the small rhyolite dome half a mile southwest of the old Keenan Ranch house (N.W. ¼ Sec. 22, T. 16 N., R. 1 E.). More detailed notes on the petrographic characters of the rhyolites and dacites are contained in the Appendix.

It should be noted, in conclusion, that there is no discernible pattern in the distribution of the dominant rhyolite domes with respect to the subordinate dacite domes, and many are juxtaposed. Close to the northeast side of the andesitic core, for example, there is a composite cluster of siliceous domes almost two miles long. Most of them are composed of biotite rhyolite, but some small ones consist of hornblende-biotite dacite. And a short distance south of this dome-cluster, close to Dean's Place, there are at least four small domes on an east-west line, the two eastern ones consisting of quartz- and biotite-rich rhyolites whereas the others consist of quartz- and biotite-poor dacites.

Exposed Intrusions of Dacite and Rhyolite

Many irregular siliceous intrusions are present within the andesitic core of the buttes, particularly in the vicinity of South Butte. All consist of whitish lava with a dense, generally unbanded porcelain-like matrix of micro- and crypto-felsite. Even within individual intrusions, the content of phenocrysts varies widely, some parts being rich in phenocrysts of biotite, quartz, and feldspar whereas others are almost devoid of them. The anomalous, approximately north-south strike of Cretaceous sediments a few hundred yards south of the largest of the exposed bodies of rhyolite probably indicates concealed extensions of the same intrusion.

A small patch of biotite-rich rhyolite outcrops among the andesites close to the western edge of the central lake-beds, and another is exposed a quarter of a mile to the west. The relations of these rhyolites to the andesites are obscure.

An elongate body of hornblende-biotite rhyodacite on the northeast edge of the buttes (Sec. 10, T. 16 N., R. 2 E.) forms a group of rocky knolls that trend toward the center of the buttes (plate 2b). A gas well drilled in line with the knolls, approximately a mile away, may have cut the same body of dacite at a depth of 1,600 feet; moreover, the axis of an elongate "structural high" on the top of the Kione beds (fig. 11) continues southwest for another mile or so. The intrusion may therefore dip toward the center of the buttes at an angle of approximately 17°.

Buried Intrusions of Rhyolite

Our next concern is with domical bodies of rhyolite that rose mostly, if not entirely, during the explosive episode though they now lie buried beneath Rampart beds. Some of these bodies only uparched Cretaceous and Tertiary beds; others rose to higher levels so as to uparch lower and middle members of the Rampart formation before being buried beneath the smooth slopes of the topmost Rampart beds. Drilling has shown that several uplifts were caused by rising bodies of rhyolite; perhaps all were formed this way, though the possibility of buried andesite intrusions cannot be denied.

Geologists of the Buttes Gas and Oil and of the Shell Oil Companies generously provided data concerning the depths at which rhyolites were cut within the Sutter Buttes. Rhyolites cut by wells in the buried "Colusa Buttes" are described in a later section (see "The Buried Colusa Buttes").

A well drilled roughly 2 miles northeast of Pennington (Sunray-Justeson Well No. 1; Sec. 21, T. 17 N., R. 2 E.) is said to have cut "biotite dacite porphyry" at a depth of 5,438 feet. Siliceous intrusions were also cut by wells in the adjacent, northern part of the buttes. One well close to Peace Valley (Shell-Buttes Community No. 1; Sec. 12, T. 16 N., R. 1 E.) began in Rampart beds and finished in basement norite at a depth of 6,561 feet, cutting several "sills of rhyolite" separated by shales at depths of approximately 6,000 feet. Another well, on the northwest flank of the buttes, near Keenan's Ranch (Richfield-Buttes Community Well D - 1; Sec. 15, T. 16 N., R. 1 E.), which cut the basement at a depth of 7,226 feet, intersected "sills of rhyolite" at depths of 5,149 to 5,200 feet, 5,710 to 5,750 feet, and 6,190 to 6,530 feet.

Rhyolitic intrusions were again cut by wells on the east flank of the buttes. A well close to the inner margin of the Rampart in Kinch Canyon (S.E. corner of Sec. 19, T. 16 N., R. 2 E.) cut rhyolite at a depth of 5,562 feet; another that began in Rampart beds was drilled about a mile north of Sutter City. This cut rhyolite at a depth of 3,900 feet. And approximately half a mile north-northwest of Sutter City, a well spudded in Rampart beds close to the West Butte Pass road, cut rhyolite at a depth of 4,735 feet (see map, fig. 4).

Rhyolites were also cut by wells on the south flank of the buttes, beneath both the sedimentary moat and encircling rampart. The map (fig. 4) shows that the largest exposed dome of rhyolite lies close to the former hamlet of West Butte. Wells drilled in the moat approximately a quarter of a mile to the south cut rhyolites at depths of 4,800 and 5,270 feet, and about a mile farther south a well begun in Rampart beds cut rhyolite at a depth of 5,996 feet. These buried rhyolites may belong to the same body as that exposed at the surface. Rhyolites were also cut by wells near the southeast edge of the buttes, at depths of 4,720 and 5,060 feet (fig. 6).

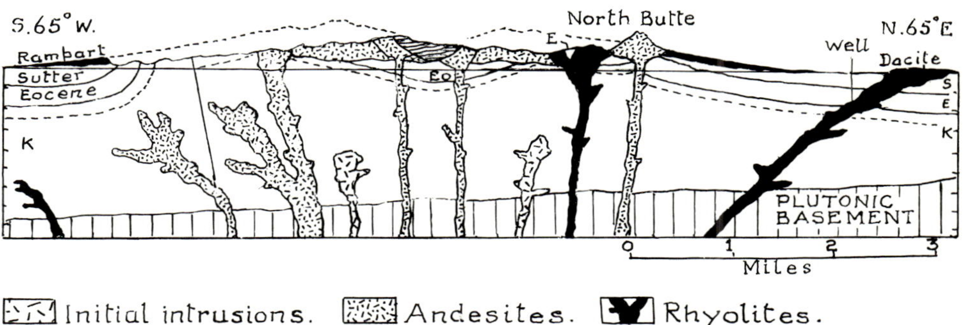

Fig. 6. Schematic sections through the Sutter Buttes: No. 1 – Richfield-Buttes Community "D" Well; No. 2 – Richfield-Buttes Community "B" Well.

A few years ago, excavations were made 2½ miles southwest of Pennington (N.W. corner of Sec. 12, T. 16 N., R. 1 E.). These penetrated a thin cover of Rampart beds resting on a channeled surface of steeply upturned white sands. Unfortunately, we were unable to determine whether these sands belong to the Ione or Kione formation, but in any case their occurrence at shallow depths denotes pronounced uparching of pre-volcanic sediments, followed by deep erosion prior to the final stages of the explosive episode. The structure-contour map (fig. 11) shows two "highs" northwest of this locality, and a gas well drilled about a quarter of a mile to the northwest is said to have cut "andesite and rhyolite" containing phenocrysts of hornblende. If, as we assume, the hornblende was identified correctly, the rock must have been either a dacite or andesite, not a rhyolite.

Rampart Beds

Introduction

How long an interval of erosion accompanied uparching of the Cretaceous and Tertiary sedimentary beds prior to the outbreak of explosive activity is not known, but when the first Rampart beds were deposited, the erosion-surface cut across the upturned sedimentary beds in the moat was not very different from that exhumed since volcanism came to an end.

The long volcanic history of the Sutter Buttes is epitomized by events that occurred during 1944 and 1945 at Syowa Sinzan in Hokkaido, Japan (Minakami *et al.*, 1951). For that reason, we present the following brief synopsis of these eruptions.

The new volcano rose in flat ground close to the base of the large composite Usu volcano. Quakes began on December 28, 1943, and continued during the early months of 1944 while the flat ground was being upheaved about 50 feet and severely fissured. More rapid upheaval and more severe fissuring began in April; and on June 22, 1944, explosions started near the center of the upheaved ground. At first, these were geyser-like mud eruptions caused when groundwater was vaporized by rising magma, but, beginning on July 1 and continuing until the end of October, there were intermittent explosions of lithic debris composed of finely comminuted chips of sedimentary rocks. All explosions were of phreatic (steam-blast) type; at no time was fresh magma discharged as incandescent ash, lapilli, and bombs. At the close of the explosive phase, there were several craterlets on top of the upheaved ground, and the new, circular "Roof Mountain" measured approximately 650 feet in height and 3,300 feet in diameter at the base. Deformation extended locally for as much as 6 miles south of Usu, and the area uplifted more than 3 feet measured no less than 5,000 by 6,000 feet across. Finally, about mid-November, 1944, viscous lava broke through the top of the "Roof Mountain." Explosions and quakes ceased, but slow extrusion of lava continued until September, 1945, by which time a steep-sided Pelean dome of dacite had been built to a height of about 500 feet, with a basal diameter of about 1,000 feet (plate 4a).

Explosive activity at the Sutter Buttes was mainly of the phreatic (steam-blast) type. Consequently, most ejecta are lithic, i.e. they were already solid when expelled. Eruptions involving discharge of fresh, pumiceous ejecta were almost entirely restricted to the initial phases of activity. Many small explosion-pipes must have been drilled by steam blasts as groundwater penetrated to rising magmas; but, as at Syowa Sinzan, none of the explosions were violent, and none ejected large blocks or bombs. All the coarse debris in the Rampart beds was derived non-explosively from fractured crusts and talus slopes of Pelean domes.

Except for a few thin lenses of gravity-sorted, airfall ejecta, all the Rampart formation consists of waterlaid volcanic sediments, derived mostly by reworking of pyroclastic ejecta that once mantled the central part of the buttes, and partly by reworking of blocky talus deposits bordering Pelean domes.

The formation may be divided for convenience into three members, though they are not sharply divided. (See figs. 7 and 8.)

1. *Basal Member.* Consists chiefly of pale-colored, fine-grained, fluviatile volcanic sediments derived from airfall deposits composed principally of pumiceous and siliceous hornblende-biotite andesite, dacite, and rhyolite admixed with a small amount of pre-volcanic, sedimentary debris picked up by streams as they traversed the sedimentary moat (plate 6b).

2. *Middle Member.* Consists almost exclusively of waterlaid andesitic lithic debris derived from steam-blast eruptions. Includes lenses of coarse laharic material carrying blocks of andesite and rhyolite, and lenses rich in pre-volcanic, sedimentary debris (plate 6a).

3. *Upper member.* Composed almost wholly of stringers of coarse, bouldery andesitic debris laid down by lahars and perhaps in small part by "block and ash flows" from the flanks of the youngest Pelean domes (e.g. those forming North, South, and West Buttes).

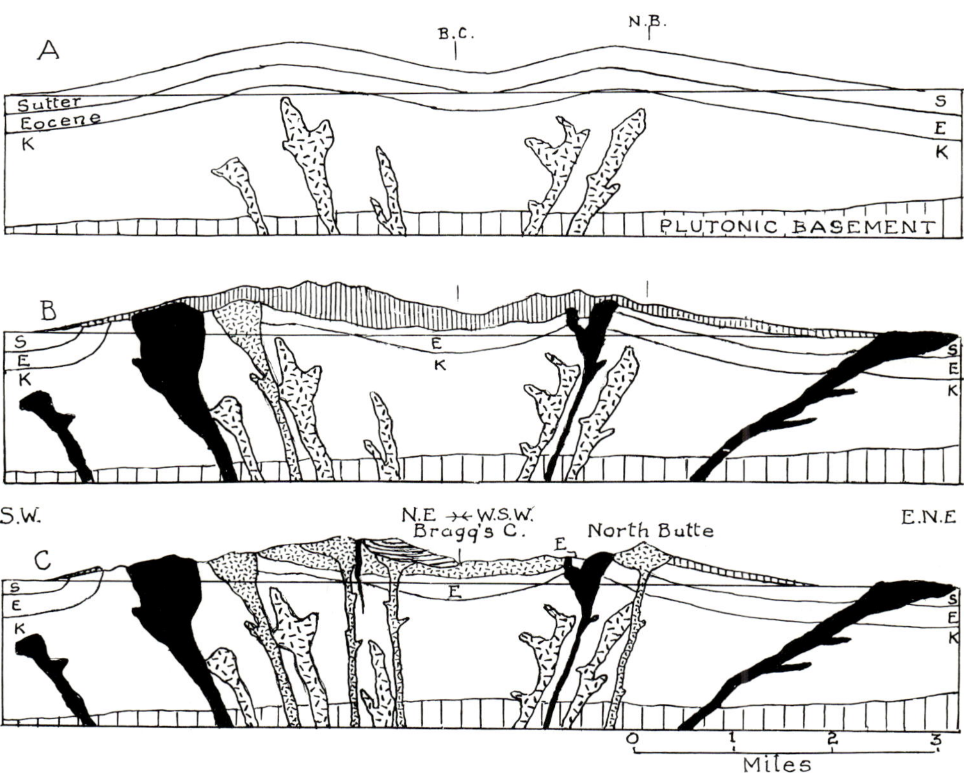

Fig. 7. Evolution of the Sutter Buttes. Idealized sections. Explanation of patterns as in figure 8. A — initial, pre-explosive uplift; effects of concomitant erosion not shown. B — Early explosive phase, before accumulation of the central lake-beds. C — Present section.

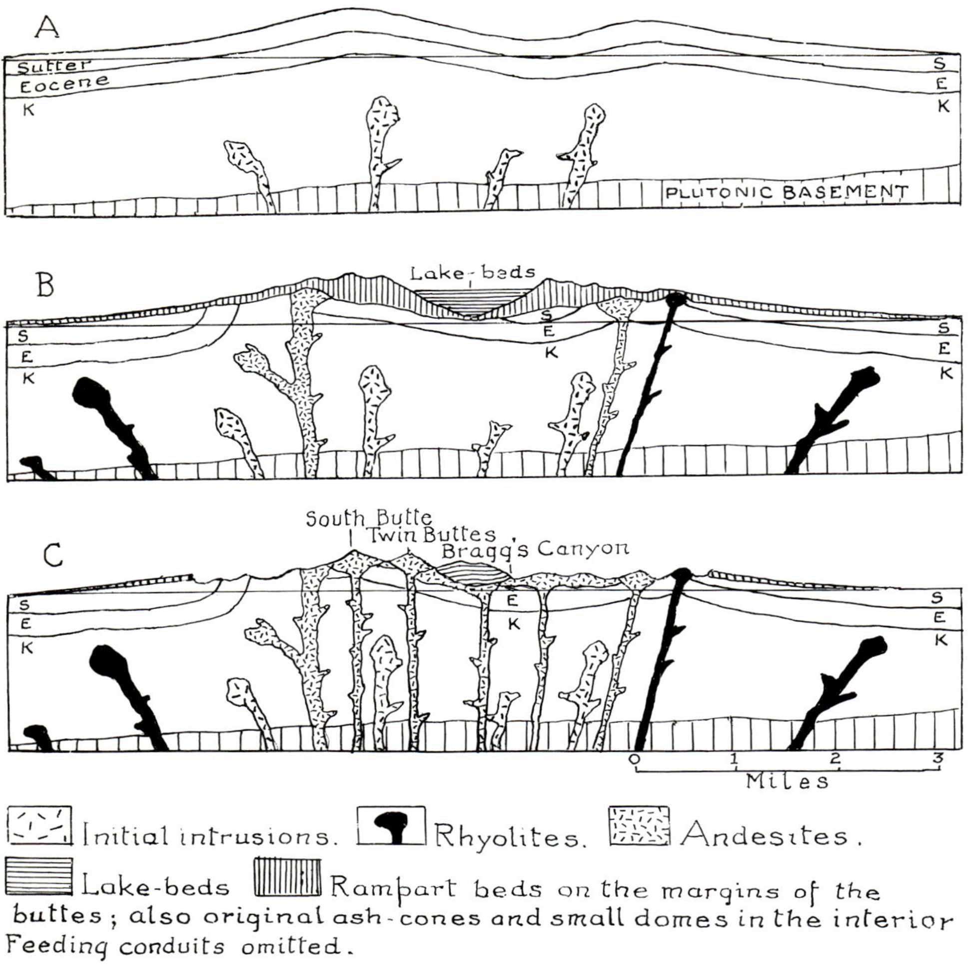

Fig. 8. Evolution of the Sutter Buttes. Idealized south-north sections. A — Initial, pre-explosive uplift. Effects of concomitant erosion not shown. B — Near close of early explosive phase, after accumulation of the central lake-beds. Many ash-cones and some small Pelean domes in the interior of the buttes, but their feeding pipes are omitted. Fluviatile Rampart beds on the flanks. C — Present section, after growth of the youngest Pelean domes of andesite. All pyroclastic deposits have been eroded from the center of the buttes except for some that may be buried beneath andesitic lavas.

Basal Member

The first products of explosive activity may have been removed by erosion and now lie buried beneath the Sacramento Valley, and presumably they consist of finely comminuted Cretaceous and Tertiary sedimentary rocks.

The oldest of the exposed Rampart beds are dominantly pale-colored tuffaceous sediments produced by reworking of pumiceous and vitric, airfall ejecta. On the inner wall of the rampart south of South Butte, the earliest beds consist almost wholly of biotite rhyolite debris; elsewhere, the basal member is composed chiefly of hornblende-biotite andesite and subordinate dacite. In some places, particularly near North Butte and in the Morehead Quarry at the south edge of the buttes, the beds include laharic layers that carry blocks of pumiceous lava; elsewhere, they include sandy and pebbly lenses rich in pre-volcanic, sedimentary debris incorporated by streams and lahars that crossed the sedimentary moat.

The basal beds outcrop at scattered localities on the south, west, and east sides of the buttes, but they are thickest near North Butte, where they reach a maximum thickness of almost 500 feet, though elsewhere they are rarely more than 200 feet thick.

All but two remnants of the basal beds are to be seen along the inner wall of the rampart. One exception is a patch at the north edge of the andesitic core of the buttes (N.E. corner of Sec. 14, T. 16 N., R. 1 E.). An account of this unusual deposit has already been given in Williams' 1929 paper (pp. 186-187). The principal outcrops lie within an oval depression a quarter of a mile long and 200 yards across, confined at one end by cliffs of massive andesite. No precisely similar deposits are known within the buttes, though they share many features with the basal Rampart beds close to North Butte and near Dean's Place at the head of Kinch Canyon. Most of the deposits were laid down by streams. Some beds, up to 10 feet thick, are unstratified and carry fragments, up to fist-size, in a matrix of rhyolitic ash. These probably represent laharic deposits laid down by floods. Other beds consist of volcanic sand and gravel; still others, only a few inches thick, consist of clay-size particles. Taken together, these finer=grained beds suggest deposition from slow-moving streams and in ponds.

The deposits within this outlier, in common with all other basal beds of the Rampart formation, are typified by abundant pale-colored rhyolitic, andesitic and dacitic debris, partly pumiceous and partly dense. Mingled with this material are hornblendic pebbles and cobbles derived from "basic inclusions" such as may be found in adjacent andesites. In addition, there are small fragments of sandstone and limestone derived from Tertiary beds that formerly covered the central part of the buttes. No other outcrops of the basal Rampart beds contain as much pre-volcanic debris. Though the deposits were laid down horizontally or almost so, they now dip at angles of between 10° and 40°, no doubt owing to uplift by adjacent bodies of andesite.

The other remnant of the basal Rampart beds preserved within the core of the buttes is cut by the trail up Brockman Canyon. It lies approximately 0.8 mile N. 60°W. from the top of South Butte. It consists of well-bedded fluviatile or lacustrine rhyolitic sediment, dominantly made up of clay- and silt-size particles. The clayey matrix is crowded with minute, angular chips of quartz, abundant flakes of fresh biotite, a few chips of plagioclase and occasional ones of sanidine. Hornblende and pyroxene are lacking.

The basal Rampart beds are particularly well-exposed at the east end of Peace Valley, near North Butte, where some sections range between 400 and 500 feet in thickness (Williams, 1929, pp. 185-186). Fragments of dense and pumiceous rhyolite, andesite, and dacite generally combine to make up more than three-quarters of the volume, the remainder consisting of angular chips of Eocene sandstone, siliceous pebbles from the Butte Gravels, bits from the Sutter Beds and occasional fragments, up to 8 inches across, of hornblende granodiorite or quartz diorite.

A mile northwest of North Butte, the basal beds are about 150 feet thick and rest on a deeply eroded surface of Tertiary sediments; indeed, at one place, they appear to rest directly on upturned Butte Gravels, the intervening Sutter beds having been removed by erosion prior to the explosive episode. They consist chiefly of well-bedded, waterlaid volcanic silts and sands rich in discrete crystals of hornblende, biotite, and plagioclase together with a few of quartz admixed in a base of clay and devitrified glass-dust. Virtually all the plagioclase crystals are splinters; hence we assume that the deposits were first laid down by airfalls during explosive eruptions and were later carried by streams to their present positions. Alignment of biotite flakes is largely responsible for their bedding and lamination.

On the opposite, southeast side of North Butte, near Dean's Place (on the boundary between Secs. 19 and 20, T. 16 N., R. 2 E.), outcrops of basal beds consist mainly of hornblende-biotite andesite or dacite and biotite rhyolite ejecta mixed with a few angular fragments of hornblende-biotite quartz diorite torn from the pre-Cretaceous basement, and occasional pebbles from the Butte Gravels.

Approximately a mile farther southeast, near the head of Kinch Canyon (close to the center of the northern boundary of Sec. 29, T. 16 N., R. 2 E.), there is a dramatic exposure of an unconformable contact between the basal Rampart beds and underlying Sutter beds, the former dipping 10° to 15° N.E., whereas the latter dip in the same direction at about 50°. Hereabouts, as might be expected, the thickness of the basal Rampart beds varies considerably, but nowhere does it exceed about 150 feet. The lower part is made up almost exclusively of fragments of biotite rhyolite, hornblende-biotite andesite and dacite; upward the content of andesitic debris increases. Scattered throughout are pebbles from the Butte Gravels and subangular fragments, up to 4 inches across, of uralitized gabbro similar to some of the basic inclusions in the andesite domes, along with a few bits of unaltered hornblende-biotite quartz diorite torn from the plutonic basement beneath the buttes.

A quarter of a mile west of the foregoing locality, there is a conspicuous mesa capped by bouldery andesitic lahars that comprise the topmost member of the Rampart formation. Some lahars rest directly on the basal member; others rest on Sutter beds, both the middle and lower members having been stripped by erosion before the laharic deposits were laid down.

A few exposures of the basal beds are to be seen near the old Keenan's Ranch house on the northwest flank of the buttes. For example, about half a mile east of the Richfield-Buttes Community D-1 well (close to the N.W. corner of Sec. 14, T. 16 N., R. 1 E.), there are cliffs, up to 60 feet high, composed of pale-colored fluviatile beds containing angular and waterworn fragments of rhyolite or dacite mingled with abundant pebbles and cobbles from the Butte Gravels. These beds, together with the underlying Sutter beds and Butte Gravels, were tilted steeply northward and then deeply channeled before the undeformed laharic beds forming the topmost member of the Rampart sequence were laid down.

Along the west side of the buttes, exposures of the basal beds are few and small, but on the south side they are widespread and much thicker. They outcrop both on the inner wall of the rampart and locally on the outer slopes where they have been uparched by concealed bodies of andesite and rhyolite (fig. 5). A typical sample, collected close to the northwest corner of Sec. 2, T. 15 N., R. 1 E., consists of waterlaid volcanic sand and silt carrying many angular, and a few rounded fragments of siliceous andesite mixed with occasional pebbles derived from Tertiary beds. The fine matrix is composed chiefly of crystals of biotite, hornblende and plagioclase in approximately equal amounts, along with fragments of devitrified, pumiceous glass.

Well-bedded, fluviatile volcanic sediments outcrop on the inner wall of the rampart south of South Butte. These may well be the oldest of the basal beds, and they differ from the others in being composed almost entirely of dense and pumiceous fragments of biotite rhyolite.

But by far the finest exposures are to be seen in the newly opened Morehead "rock-and-gravel pit" at the extreme southern edge of the buttes, where both the basal and middle members of the Rampart formation are present. The basal member consists of fluviatile and laharic beds composed almost entirely of pale-colored, pumiceous, hornblende-biotite andesite. Much of the fine material represents reworked crystal-vitric ash, but the large blocks, many of which measure several feet across, were derived from talus surrounding Pelean domes. Mingled with the volcanic debris, but generally concentrated in thin layers, are occasional concretions of limey sandstone derived from Cretaceous beds, and more numerous pebbles derived from the Butte Gravels. These were incorporated by streams as they crossed upturned beds in the moat. In the overlying Middle Member, the content of reworked andesitic, lithic debris increases, along with the number of large, angular blocks derived from Pelean domes.

A small patch of the basal member lies in a deep channel cut in vertical Cretaceous beds within the moat, about a mile south-southwest of South Butte. This consists mainly of small chips of biotite rhyolite mingled with pebbles from the Butte Gravels, in a fluviatile matrix.

Whereas the basal Rampart beds south of the West Butte Pass road generally dip southward at about $10°-20°$, the beds in this small outlier dip in the same direction at angles of approximately $30°$, indicating more pronounced uplift close to the interior of the buttes.

Middle Member

The middle member of the Rampart formation constitutes by far the major part and over large areas it is the only member. Its thickness generally diminishes toward the margin of the buttes, but nothing is known about its thickness or extent beyond the margin. Some beds were laid down on deeply dissected surfaces and some were partly or wholly removed by erosion following local intra-volcanic uplifts; hence the thickness of the middle member varies greatly even over short distances. Adjoining the West Butte Pass road, for example, there are places where coarse laharic deposits of the upper member transgress both the middle and lower members so as to rest directly on Sutter beds. In general, however, the thickness of the middle member on the inner wall of the rampart ranges from about 250 to 500 feet, reaching a maximum on the north side of the buttes, on the wall overlooking Peace Valley.

The deposits of the middle member were laid down almost entirely by streams, most of which flowed gently though some were swift, torrential lahars carrying large blocks, and commonly the coarse materials occupy channels cut in the finer sediments. Beds composed of airfall ejecta are exceptional, and all form small, thin lenses. Nowhere are there any impact pits, such as are made by falling bombs and blocks; in a few places, however, the fine-grained airfall deposits exhibit graded bedding.

The volcanic sediments are mostly well-stratified, their constituents generally ranging in coarseness from sand- to gravel-size. Layers made up principally of silt- and clay-size particles are rare, but unstratified and obscurely stratified laharic layers are common. These are mostly between 5 and 15 feet thick and may contain blocks up to 6 feet across. All but a few blocks are andesitic; locally, however, there are rhyolitic ones that were incorporated by andesitic lahars as they swept across talus slopes on the flanks of Pelean domes (e.g. adjacent to the large rhyolite domes in Sec. 33, T. 16 N., R. 1 E.). But no matter whether individual beds are coarse- or fine-grained, few fragments were rounded during transport, presumably because the distances they moved were short and the rate of deposition was rapid.

The composition of the volcanic debris ranges from rhyolitic, through dacitic to andesitic, and although the content of siliceous debris is relatively small and generally diminishes upward, there are marked variations both vertically and laterally. Some dacitic debris on the north side of the buttes contains phenocrysts of both hornblende and biotite, exposed domes of which are rare.

Our conclusion is that almost all of the volcanic materials in the middle member of the Rampart formation were waterlaid and derived mainly by erosion of the products of phreatic eruptions and partly by erosion of debris from the flanks of Pelean domes. A little pumiceous dacitic and andesitic material was derived by reworking of airfall deposits of truly magmatic eruptions, but nowhere are there any almond-shaped or ropy bombs or lapilli.

We call attention finally to a somewhat enigmatic strip of rocks deep within the core of the buttes, trending north-northwest, about a quarter of a mile west of the central lake-beds (map, fig. 10). Williams (1929) formerly regarded this as a pillar of andesitic lava within a central crater, but it is now thought to be an outlier of andesitic lava underlain by pyroclastic beds that are probably age-equivalents of beds in the middle member of the Rampart formation. The southern part of the outlier consists of poorly bedded lapilli tuffs and tuff-breccias composed chiefly of fragments of whitish, semi-pumiceous hornblende-biotite dacite mingled with darker, denser fragments of hornblende-biotite andesite. Close to intrusive contacts, some beds are upturned almost to verticality, and locally so intensely sheared that many pumiceous lapilli have been drawn out into slender pencils. Overlying these pyroclastic deposits is a flow of hornblende-biotite andesite, the basal part of which is strongly auto-brecciated. Upward, the lava becomes brownish and reddish owing to the presence of fumarolic hematite, and grades into massive, unbrecciated lava among the summit-pinnacles on the divide overlooking Bragg's Canyon.

Upper Member

The upper member of the Rampart formation is characterized by many large, angular blocks of andesite, and consists almost exclusively of coarse laharic deposits. Williams, in his 1929 paper, wrongly supposed that much of this debris was discharged by violent explosions; it is now apparent however, that almost all of it was laid down by lahars and avalanches that swept down from the steep flanks of the youngest Pelean domes, e.g. from North, South, and West Buttes. Some lahars and avalanches may have been hot, like the "block-and-ash flows" discharged from the flanks of the 1929-1932 domes of Mont Pelé in the West Indies (Perret, 1937), or those accompanying growth of the dome of Mount Lamington in New Guinea (Taylor, 1958).

None of the deposits, unlike some in the underlying members, have been uparched by rising bodies of magma. For this reason, they form gentle, uniform, quaquaversal slopes that conceal all but a few signs of underlying deformation.

Debris comprising the upper member consists almost wholly of porphyritic, non-vesicular hornblende-biotite andesite, all or all but a little of which was already solid during emplacement. No blocks exhibit the porous, diktytaxitic textures typical of most laharic and avalanche-deposits related to Pelean domes in the West Indies and other volcanic fields. Nor was deposition preceded by explosive, ash-forming eruptions such as are commonly related to the rise of Pelean domes elsewhere. None of the block-rich beds were laid down by glowing avalanches, nor were any of them products of base-surges, i.e. downward blasts accompanying vertical eruption-columns. On the contrary, all were laid down by cool or cold avalanches and lahars that rushed downslope from the steep sides of growing andesite domes in the interior of the buttes. The relative scarcity of similar deposits among underlying members of the Rampart formation suggests that it was chiefly during final stages of volcanism that high Pelean domes developed.

Erosion has reduced remnants of the upper Rampart beds to irregular, ill-defined trains of giant boulders, none of which are more spectacular than those that rushed down from the flanks of the North Butte dome. On the slopes to the east, there are scattered boulders of andesite up to 30 by 15 by 12 feet across, and on the slopes to the north (in Secs. 5 and 8, T. 16 N., R. 2 E.), there are trains of blocks including many more than 6 feet across and one that measures 33 by 15 by 6 feet across (plate 6c). Other block-rich lahars and avalanches swept westward from Pelean domes west of North Butte, and one of them (in. N.E. corner of Sec. 16, T. 16 N., R. 1 E.) contains a block measuring 30 by 30 by 15 feet across. Several lahars that swept westward for at least 2 miles from the West Butte dome carried blocks up to 10 feet across.

The fine matrix between the blocks consists almost wholly of comminuted andesite, but fragments of rhyolite and dacite, particularly pumiceous ones, such as are common in the underlying members of the Rampart formation, are extremely rare. In a few places, however, the upper member contains occasional blocks of dense rhyolite, no doubt swept from talus slopes surrounding rhyolite domes.

Nowhere did we detect any debris that we could identify as having come from pre-volcanic sedimentary formations except for a few pebbles and cobbles from the Butte Gravels. Other pre-volcanic sedimentary debris may be present, but too finely comminuted for detection with a hand-lens.

Reference is made in conclusion to estimates reported by Williams in 1929 concerning the temperatures of the eruptions involved in deposition of the upper member of the Rampart formation. Many andesite blocks have thin, reddish crusts and pale gray cores. Phenocrysts of hornblende from the fresh cores are greenish, and those of biotite are brownish; within the thin rinds of many blocks, however, the colors of both the hornblende and biotite crystals have been changed to russets and reds, presumably by oxidation and reheating. Experiments made to duplicate these color changes suggested that the reddened oxyhornblendes and oxybiotites could never have been subjected to temperatures of more than 600°C. And it was argued that none of the ejecta could have been reheated above that temperature. Now, however, we no longer think that the blocky debris was discharged at high temperatures; on the contrary, the green hornblendes and brown biotites of the original andesites were oxidized and reddened by fumarolic gases rising along cracks within the solidified crusts of growing andesite domes. Most avalanches and lahars that transported the giant blocks were relatively cool, if not cold. Nowhere do the deposits contain charred wood or leaves.

The De Witt Volcano

This name is given to the faulted and eroded remnants of a small andesitic volcano built on top of the Rampart beds near De Witt's Ranch on the southeast flank of the buttes (chiefly in Sec. 6, T. 15 N., R. 2 E. and Sec. 1, T. 15 N., R. 1 E.) (see fig. 9). It is probably the youngest volcano in the Sutter Buttes, unless the andesite dome of North Butte is younger.

Remnants of the volcano are scattered within a roughly circular area a little more than a mile across. The underlying Rampart and Sutter beds had already been uparched, faulted, and deeply eroded before the volcano began to grow. Hence, the erupted lavas rest on all three members of the Rampart formation and on Sutter beds. The structure-contour map (fig. 11) suggests that the volcano lies near the intersection of several faults.

The original volcano was a flattish cone, and if it ever had a summit-crater, all signs have been destroyed. From the highest point, at an elevation of 727 feet, lava poured southward to about 400 feet, and southeastward to about 100 feet. Northeastward, flows descended to a small outlier adjoining the West Butte Pass road, at an elevation of about 300 feet. A large, wide valley already existed hereabouts when the De Witt volcano was active.

Few lavas are more than 100 feet thick, and most are only about half as thick. Their total volume was probably not more than about 0.01 cubic mile. They are composed of strongly porphyritic hornblende-biotite andesite closely resembling that of most Pelean domes in the interior of the buttes. They are distinctive, however, because they are intensely auto-brecciated, so much so that in many outcrops it is difficult or impossible to distinguish them from deposits of blocky andesitic lahars. Careful scrutiny will show that the lahar-like materials grade into others in which the brecciated lava present a jigsaw appearance, adjacent fragments fitting closely together. And locally the auto-brecciated lavas grade upward into massive, almost completely unbrecciated material. On the other hand, where brecciation is most pronounced, the abundant matrix between the angular blocks is reduced to sand-size particles.

Where the lava rests on upturned Sutter beds at the south edge of the northernmost of the three chief outliers, it has been bleached and slightly altered, presumably by fumarolic or hydrothermal activity.

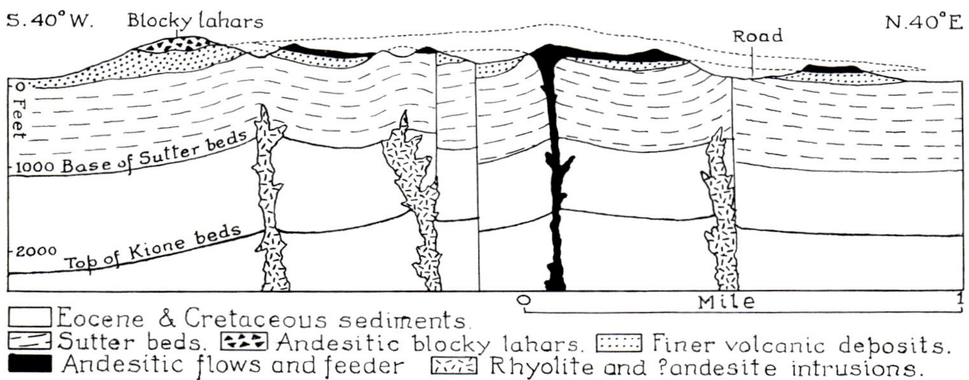

Fig. 9. Section across the De Witt volcano. The plutonic basement lies at about 7,800 feet beneath the western end of the section and 6,500 feet beneath the eastern end.

Volume of the Rampart Beds

Williams (1929) estimated that the volume of fragmental debris discharged at the Sutter Buttes was about 4 cubic miles, assuming the original cone to have risen to a height of approximately 5,000 feet. These figures must be drastically reduced, because they were based on the false idea that the present gentle slopes of the rampart formerly steepened inward to about 30°. This erroneous impression stemmed from the view that most deposits were airfall-products of explosive eruptions. Now, however, we know that almost all the Rampart beds were laid down by streams and torrential lahars. This being so, the surficial slopes at the close of the explosive episode can only have steepened slightly toward the middle of the buttes. All the major Pelean domes of andesite, such as North, South, and West Buttes, Old Craggy, and Twin Peaks, protruded well above it.

Instead of a huge, single, central cone rising to a height of 5,000 feet, as Williams (1929) and Lindgren (1895) once envisaged, there was a cluster of small, short-lived cones and domes built over many vents in the interior of the buttes, the largest of which never rose much more than 2,500 feet above sea-level. Moreover, all these features had been removed by erosion before the last Pelean domes began to grow. The gentle, outward-dipping slopes of fluviatile Rampart beds may never have risen to heights of more than 1,200 feet; indeed, the eastern slopes abut sharply against the steep flank of the North Butte dome at elevations of about 1,000 feet (figs. 6 and 7).

An accurate estimate of the original volume of fragmental material is impossible because nothing is known concerning the amount of Rampart material buried under the Sacramento Valley beyond the margins of the Sutter Buttes, and we can do no more than speculate as to the amount of airfall material that fell close to the eruptive vents to build the cone-cluster in the center of the buttes. It seems likely, however, that the total volume of reworked, waterlaid material does not exceed 2 cubic miles, and it may be considerably less.

Central Lake-Beds

There is a roughly oval area in the center of the buttes, measuring about 1.0 by 0.7 miles across, occupied by a thick pile composed partly of fluviatile, but mainly of well-bedded lacustrine, volcanic sediments. Exposures are particularly good along the south wall of Bragg's Canyon where the beds are approximately 1,000 feet thick (fig. 10 and plate 5b) and consist of sand- and gravel-size volcanic debris; there are, however, many lenticular beds of fine debris, a few composed principally of silt- and clay-size particles. Coarse, blocky layers, such as typify the top member of the Rampart formation, are notably absent. There are no layers of vitric, airfall ash and no bombs, fusiform lapilli, or other fresh magmatic ejecta. Most fragments are angular or subangular even though virtually all were waterlaid. Our impression is that the sediments consist almost exclusively of lithic debris blown out in a solid condition by weak phreatic (steam-blast) eruptions and subsequently washed into a slowly subsiding lake-basin. Not even the largest blocks produced "impact sags," as they must have done had they fallen from the air into water-soaked sediments; no doubt, therefore, all the large fragments were washed slowly into place. (See fig. 10.)

Cross-bedding, channeling, and other signs of fluviatile deposition are scarce, even close to the margins of the central basin. Stratification generally becomes more pronounced upward until, near the top, some beds are finely laminated. There is, however, no systematic variation in coarseness, either vertically or laterally. And the absence of coarse talus and laharic deposits close to the margins of the basin indicates that there were no high and active Pelean domes in the vicinity while the observable sediments were being laid down. Deposition took place in an area of low relief much closer to sea-level than at present. And we suppose that the lake-basin occupied a large depression on the summit of the original com-

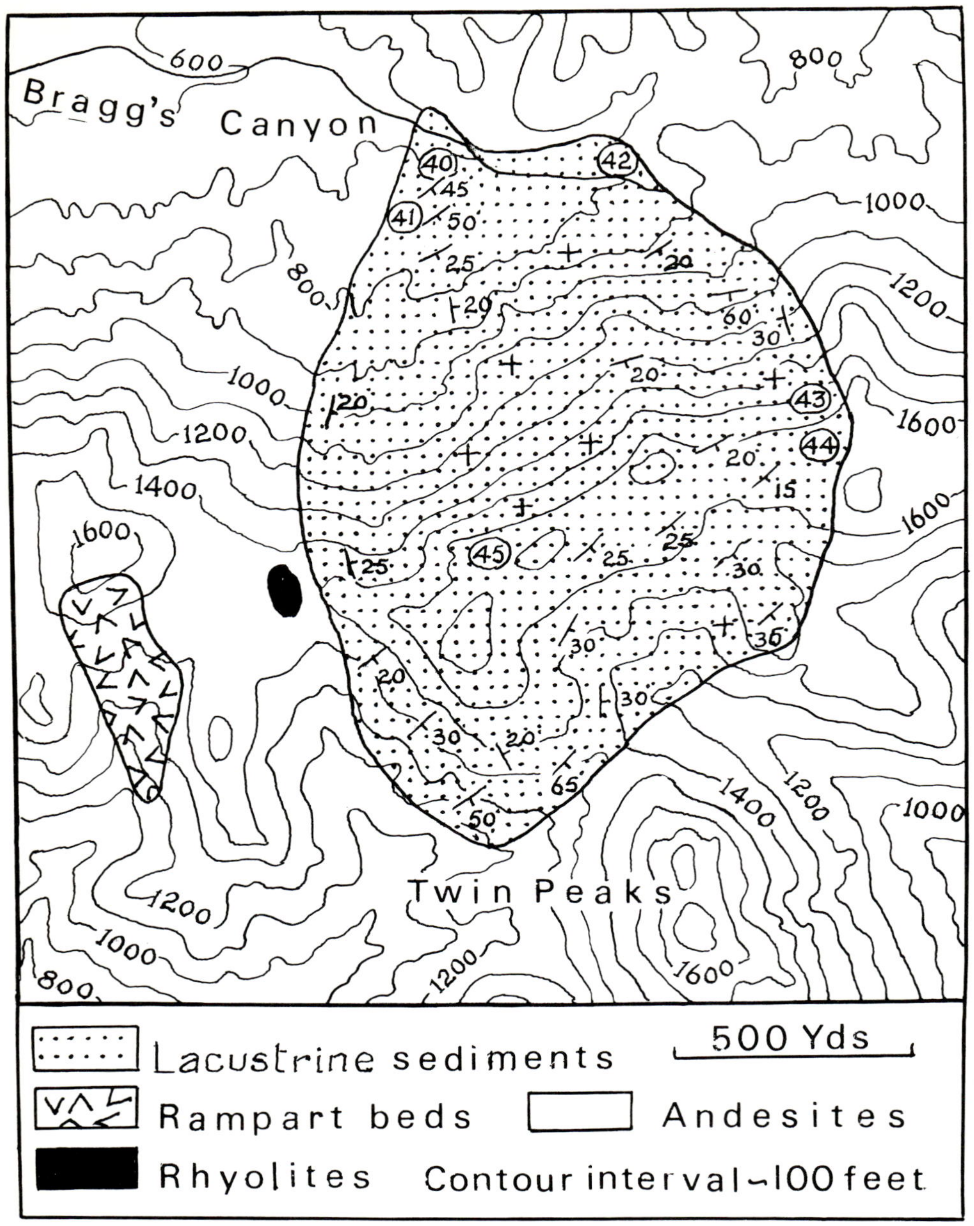

Fig. 10. Map of the central lake-beds. Specimen-numbers in circles.

posite intrusion from which the Tertiary beds had been largely stripped prior to the start of explosive activity. The present east-west Bragg's Canyon may reflect this pre-explosive depression.

As for the petrographic character of the deposits, we emphasize at once that we detected no debris derived from pre-volcanic sedimentary rocks. Most fragments consist of hornblende-biotite andesite, though some consist of biotite rhyodacite. If there is any regular variation in bulk-composition from top to bottom, we failed to recognize it. Fragments of hornblende-biotite andesite recur throughout the sequence, but it is mainly within the topmost deposits that the hornblende is fresh; elsewhere it is altered to chlorite, calcite, and "clay," suggesting derivation from propylitized andesites. Biotite flakes, on the contrary, are invariably fresh, even in the most altered deposits.

Can the lake-beds be correlated with any Rampart beds? Certainly, there are no equivalents of the blocky laharic deposits that typify the top member of the Rampart sequence; and if any of the pale pumiceous sediments that typify the lower part of the basal member are present, they must be buried under Bragg's Canyon. Our conclusion is that the lake-beds most closely resemble deposits in the upper part of the basal member and lower part of the middle member. And because they lack blocky layers, the lake cannot have been bordered by active Pelean domes as the sediments accumulated.

Consider next the attitudes of the beds. Most of the sediments must have been laid down horizontally or almost so, but most were later upheaved and deformed by adjacent and subjacent bodies of andesite. Near the center of the basin, on the south wall of Bragg's Canyon, most beds still lie horizontally or dip at angles of only a few degrees (fig. 10). Near the margins of the basin, on the contrary, many beds dip inward at moderate angles and some do so at angles of approximately 60°. And throughout the southern part of the central basin, the prevailing dips are outward, to the south and southeast at angles of 15° to 65°, the beds abutting abruptly against adjacent domes of andesite.

In his 1929 paper, Williams showed what he thought was a prong projecting southeastward from the supposed "central vent." This prong actually consists of cavernous-weathering, locally auto-brecciated, even mylonitized andesite (see Appendix), cut by thin stringers of dense, dark andesite, and it ends sharply against the well-bedded deposits of the central basin. Similar cavernous-weathering andesite can be seen on the north wall of Bragg's Canyon, close to the margin of the central basin. The caverns lie horizontally, and may be related to former levels of the central lake when some of the peripheral domes of andesite rose to the surface.

Intra-Volcanic Deformation

No volcanic region that we know of exhibits stronger or more extensive deformation of overlying beds by rising magmas than can be seen in the Sutter Buttes. Prior to the explosive episode, regional uparching of Cretaceous and Tertiary beds affected a circular area about 10 miles across, and in places beds approximately 5,000 feet thick were upturned. Our concern here is with the continued but more localized deformation during the explosive phase that produced the Rampart beds. This later deformation was caused chiefly by rising bodies of rhyolite, some of which broke through to the surface while other solidified underground. And although some deformation may have taken place shortly after the preexplosive uplift, most of it occurred when the surface cut across the Cretaceous and Tertiary beds within the sedimentary moat was not markedly different from what it is today.

Harry Johnson (1943) was the first to recognize intra-volcanic deformation of the Rampart beds. Williams (1929) had previously noted anomalous dips in several places within the Rampart beds, but wrongly assumed that these were initial sedimentary dips reflecting accumulation of airfall-ejecta on valleysides.

All the intra-volcanic deformation took place before the topmost laharic beds of the

Rampart formation were laid down. Hence the uniform, gentle quaquaversal slopes formed by these topmost beds belie the complex structures underground.

Our account begins by describing intra-volcanic deformation around exposed domes of rhyolite and above buried intrusions within the sedimentary moat, and continues by describing deformation caused by buried intrusions beneath the surrounding rampart.

Deformation Within the Moat

Evidence is particularly clear on the southwest, south, and southeast sides of the buttes. On the southwest side, the sedimentary beds once dipped in that direction as a result of the initial, pre-explosive uplift; later, during the explosive phase, a large dome of rhyolite was extruded, and talus blocks from its flanks were incorporated in adjacent andesitic lahars. The rise of this dome uparched the surrounding beds, locally reversing their earlier dips. Moreover, the exposed dome may be merely the top of a much larger, buried body of rhyolite. Refernce to figure 5 will show that a narrow, steep-sided anticline extends southeastward from the margin of the exposed rhyolite to the Brockman Canyon fault. Almost surely this sharp flexure reflects a buried extension of the exposed rhyolite. Two gas wells drilled about half a mile south of the exposed dome cut rhyolites at depths of approximately a mile. And in the area between the wells, the Sutter beds and overlying basal and middle members of the Rampart formation were strongly uparched and deeply eroded prior to deposition of the topmost laharic deposits. Presumably this strong intra-volcano disturbance was caused by buried extensions of the exposed dome of rhyolite. Other buried extensions must account for the steep dips of the sediments west of the dome where they are overlain with strong unconformity by Rampart beds (plate 3b).

Scattered outcrops along the south side of the West Butte Pass road reveal the basal member of the Rampart formation resting conformably, or almost so, on Sutter beds; hence both must have been tilted southward together during the explosive episode. Other outcrops reveal a marked angular break between the middle member of the Rampart formation and the Sutter beds, denoting renewed intra-volcanic uplift. Close to the De Witt volcano, for example, a small patch of sediment from near the base of the middle member rests against the wall of a small, steep-sided valley cut in Sutter beds.

Intra-volcanic deformation is nowhere more strikingly revealed than around the rhyolite dome that forms the conspicuous "777 Hill," close to the West Butte Pass and roughly half a mile north of the De Witt volcano. Before this dome rose to the surface, the adjacent sedimentary beds already stood almost vertically, trending approximately east-west; now, the beds swerve around the dome, almost as if it were a knot in wood (fig. 5). Close to the southern edge of the dome, pebbly Kione beds are slightly overturned; to the north and northwest, aberrant dips in other Cretaceous beds suggest buried extensions of the rhyolite in those directions. Noteworthy, also, is the fact that the "777 Hill" rhyolite, the center of the De Witt volcano, and the rhyolites cut by deep wells farther south lie approximately on a line which is radial with respect to the center of the buttes.

The elongate rhyolite dome on De Witt's Ranch, on the southeast flank of the buttes, caused relatively little deformation of adjacent Sutter beds. It must have risen to the surface at an early stage in the explosive phase, because talus blocks from its western side are incorporated in adjacent Rampart beds (see plate 13a of Williams' 1929 paper). A short distance to the north, in line with this elongate dome, is a small, circular one, probably of the same age, which strongly uparched the surrounding Butte Gravels. And a buried extension of this dome may have been cut by drilling half a mile to the east, through the floor of Kinch Canyon.

On the northeast side of the buttes, close to North Butte, the sedimentary moat is covered by Rampart beds, and if the exposed domes of rhyolite and rhyodacite uparched the sedimentary beds, the evidence is concealed. However, the small rhyolite dome that rises from Peace Valley distinctly uparched adjacent Butte Gravels.

Deformation Within and Beneath the Rampart

Although the uniform, gently dipping outer slopes of the topmost Rampart beds almost completely conceal the extensive deformation of underlying beds, surface mapping and geophysical measurements clearly indicate many buried uplifts. One such uplift underlies the De Witt volcano, and almost surely all the others are related to buried intrusions, chiefly of rhyolite. (See fig. 11.)

Consider first the uplift related to the De Witt volcano. This volcano, perhaps the youngest in the buttes, rests on uparched and faulted Sutter and Rampart beds, and minor faulting continued after the volcano became extinct. The map (fig. 5) and section (fig. 9) indicate the general structure. A roughly east-trending fault, with small upthrow on the south, is traceable for about two miles along the inner slope of the rampart from the Brockman Canyon radial fault to the De Witt volcano where it swings southward, bifurcating into two branches that reunite about a mile from the volcano. A well drilled between the branches of the fault cut rhyolite at a depth of 4,720 feet; another cut rhyolite, a mile farther south, at a depth of about 5,060 feet. The structural uplift beneath the De Witt volcano may have been caused by a buried body of rhyolite; more likely, however, it was caused by the body of andesite that fed the volcano.

The Sutter Buttes gas field lies west of the De Witt volcano (in Secs. 1 and 2, T. 15 N., R. 2 E.). Thamer (1961) says that production here is not controlled by structure but by lenticular sands within the Cretaceous Forbes shale. It should be noted, however, that the field lies within a faulted and broadly warped, south-dipping monocline which is almost surely of volcanic origin.

The Sutter City gas field, between the De Witt volcano and Sutter City (Secs. 7, 8, 17, and 18), is said by Thamer to lie within a southwest-plunging anticline sealed updip by a northwest-trending fault. It is bordered on the southeast by a northeast-trending fault with a downthrow of about 150 feet to the southeast. Production comes from Kione sands at depths of approximately 2,500 feet (see structure-contour map, fig. 11). This plunging anticline and associated faults almost certainly reflect buried intrusions, probably of rhyolite.

Buried bodies of rhyolite were cut at depths of 3,900 and 4,735 feet by wells a short distance north of the Sutter City gas field. No doubt these account for the strong domical uplift of the Kione beds depicted on the structure-contour map (fig. 11).

On the east side of the buttes, near the head of Kinch Canyon, the topmost Rampart beds occupy channels up to 150 feet deep cut in beds belonging to the middle member, and in places they rest on Sutter beds, providing eloquent testimony to strong intra-volcanic uplift and erosion. A well drilled through the floor of the canyon (where it passes from Sec. 29 to 28, T. 16 N., R. 2 E.), on the nose of an asymmetrical, east-plunging anticline, cut rhyolite at a depth of 5,562 feet. This rhyolite is probably an offshoot from the one that rises to the surface half a mile to the west, and we think that the buried intrusion accounts for the intra-volcanic deformation just described.

Approximately a mile north of the head of Kinch Canyon, close to Dean's Place, the basal member of the Rampart formation dips to the east and northeast at angles ranging from 15° to 70°. Buried extensions of adjacent siliceous domes are probably responsible for this deformation.

A deeply buried body of andesite or dacite is present under the northern edge of the rampart, roughly 2.5 miles southwest of Pennington. This intrusion may be merely one of a northwest-trending series that produced the domical uplifts shown on the structure-contour map (fig. 11).

Evidence for both pre- and intra-volcanic deformation on the northwest flank of the buttes, near the old Keenan Ranch house, has already been cited. To this evidence must be added that to be seen about a mile west of the ranch-house (N.E. ¼ of Sec. 16, T. 16 N., R. 1 E.), where a deep north-south gully cuts the Rampart beds. Close to the mouth of the gully, dips are to the north at about 10°; higher in the section, upper beds of the middle member and the overlying laharic beds of the top member generally dip westward at only 5°.

Intra-explosive deformation along the southwest side of the buttes was produced by a large cluster of rhyolite bodies, only one of which is exposed. The unusually steep south-westward dips of the Kione beds in this area is shown on the structure-contour map (fig. 11).

The foregoing examples of intra-volcanic deformation were probably produced over an extensive period, and most uplifts were accompanied and followed by such rapid erosion that the final deposits of the Rampart sequence completely buried them beneath uniformly dipping slopes.

The Buried "Colusa Buttes"

Petroleum geologists have given the name "Colusa Buttes" to a large uplift beneath the floor of the Sacramento Valley close to the village of Colusa, approximately 4 miles west of the Sutter Buttes. The uplift is roughly oval, measuring about 12 miles in a north-south direction and between 3 and 4 miles across. Within the buried buttes, Cretaceous beds have been uparched locally to heights of more than 1,500 feet (fig. 11). This impressive uplift, like that of the sedimentary beds within the Sutter Buttes, was caused by the rise of extremely viscous bodies of magma, and almost surely all uplifts took place at about the same time.

Whether or not the uplifts in the "Colusa Buttes" were caused in part by the rise of andesitic magma is uncertain, but at least six gas wells cut porphyritic rhyolites at depths ranging between 5,600 and 9,470 feet. Available information concerning these wells is as follows.

Humble O. P. Davis B-2 Well (Sec. 26, T. 15 N., R. 1 W.) cut two "sills" of rhyolite intrusive into Cretaceous shales at depths between 9,400 and 9,470 feet. A sample from 9,430 feet (Univ. Calif. Coll. 134-104) is a biotite-rich rhyolite similar to many rhyolites exposed within the Sutter Buttes. A tenth consists of sub-parallel flakes of biotite, up to 1 mm. across, characterized by rims that are pleochroic from pale yellow to deep russet and by much darker cores pleochroic from olive-green and brown to blackish brown. Euhedral and slightly embayed bipyramids of quartz, one of which measures 4 mm. across, constitute roughly 5 percent of the bulk. Euhedral, pseudo-uniaxial phenocrysts of sanidine make up an equal amount, while zoned phenocrysts of oligoclase make up about 4 percent. In addition to several minute zircons, there are many stumpy prisms of apatite. Most of the rhyolite appears white and porcelanous in hand specimens, consisting of cloudy "cryptofelsite" crowded with slender microliths of sanidine and oligoclase. We think it is not unlikely that this extremely fine-grained groundmass was originally glassy. An almost identical lava, differing chiefly in its lower content of biotite, was cut by the nearby Amerada Company well.

Humble O. P. Davis B-5 Well cut "rhyolite porphyry" at a depth of 9,300 feet, while the B-3 Well cut "sills" of similar material at various depths between 9,300 and 9,800 feet. The Atlantic Franco-Western Steidlmeyer Well (Sec. 2, T. 15 N., R. 1 W.) is reported to have cut "creamy white granitics with traces of mica" at depths of between 5,600 and 5,800, 5,900 and 6,050 feet, again at 6,800 feet. We have not seen samples of these so-called "granitic" rocks, but we feel confident that they are biotite rhyolites.

Our conclusion is that the buried "Colusa Buttes" were uparched principally, if not exclusively, by rising bodies of unusually viscous rhyolite. And we emphasize that although these rhyolites were intruded at depths of approximately a mile to 2 miles, they are texturally and mineralogically indistinguishable from rhyolites that rose to the surface within the Sutter Buttes. To be more specific, the rhyolite cut by the Humble O. P. Davis B-2 Well at a depth of 9,430 feet is almost identical in thin-section to the rhyolite exposed near Keenan's Ranch within the Sutter Buttes (Univ. Calif. Coll; slide 16, new series). Both have micro- to crypto-felsitic matrices that may once have been glassy. The deep-seated, but fine-grained rhyolites in the "Colusa Buttes" must have been suddenly and drastically chilled, presumably by injection into cold, water-soaked sediments. Their exceptionally high viscosity helped them to produce intense deformation of the adjacent sedimentary beds.

The Wild Goose gas field, about 7 miles northwest of the Sutter Buttes (Secs. 17 and 18, T. 17 N., R. 1 E.) occupies an elongate dome, roughly a mile by a mile-and-a-half across. Humble Wild Goose Well No. 6 cut a basement of hornblende-rich diorite at a depth of 6,697 feet; nevertheless, we agree with Hawley (1962) that the domical uplift was probably caused by another buried Quaternary intrusion, though whether andesitic or rhyolitic is not known.

We note finally that a well drilled 4 miles northeast of the Sutter Buttes and roughly 3 miles west of Gridley cut "biotite dacite porphyry" at a depth of 5,269 feet. We have not seen a specimen, but suggest that the lava may be similar to and of about the same age as the dacite exposed close to the northeast edge of the buttes.

Age of Volcanism

Precisely when magmas began to uparch the sedimentary beds in the area now occupied by the Sutter Buttes is not known. There can be little doubt, however, that the initial uplift took place long before the first explosive activity. When the first explosions occurred is also uncertain, because their ejecta seem to have been removed by erosion and redeposited in the Sacramento Valley, beyond the limits of the buttes, before the oldest of the exposed deposits of the Rampart formation were laid down. Long chapters in the early volcanic history are missing.

According to potassium-argon determinations, the exposed volcanic rocks range in age from about 2.5 m.y. to 1.4 m.y. We admit, however, that some age-determinations are difficult to reconcile with observations made in the field. The absence of signs of deep erosion and soil-horizons suggests that accumulation of the Rampart beds was essentially a continuous process, without long intervals of volcanic quiet. Accordingly, we think that the basal beds cannot be much older than the first domes in the core of the buttes; in other words, the first explosions probably took place about 2 million years ago.

The large andesitic dome of North Butte and the andesitic flows of the De Witt volcano seem to be products of the last eruptions despite the radiometric results. Moreover, it seems unlikely that volcanism continued at short intervals for as long as 1.5 m.y.; even the largest volcanoes in the Cascade Range and the giant Mauna Loa in Hawaii were built in much less than a million years. The entire explosive episode and rise of Pelean domes in the Sutter Buttes probably took place within approximately half a million years, from about 2.0 to 1.5 m.y. ago.

There is no regular sequence in the types of erupted magmas; on the contrary, andesites, dacites, and rhyolites were erupted in irregular order, and some may have been discharged simultaneously.

The following table presents the radiometrically dated rocks in order of diminishing age.

Radiometric Ages   M.Y.

1. *Rhyolitic fragments from the basal Rampart beds.* South of the West Butte Pass road; close to the middle of the north boundary of Sec. 2, T. 15 N., R. 1 E. Sample No. 62.   **2.4**

2. *Biotite rhyolite* from the buried "Colusa Buttes" Depth, 9,430 feet. Sec. 26, T. 15 N., R. 1 W.; Humble Well.   **2.2**

3. *Hornblende-biotite andesite* from dome in the moat on the northwest side of the buttes. N.E. corner of Sec. 21, T. 16 N., R. 1 E. Sample No. 64.   **2.0**

4. *Rhyolite dome* on the east margin of the andesitic core; near center of Sec. 30, T. 16 N., R. 2 E. Sample No. 16. Illustrated in plate 18b and c of Williams' 1929 paper.   **1.89**

5. *Hornblende-biotite andesite* from the core of the buttes, approximately ½ miles east of the summit of West Butte.   **1.8**

6. *Hornblende-biotite andesite boulder* in the coarse laharic deposits derived from the flank of North Butte. Three miles N.N.E. of North Butte, and 2 miles S.E. of Pennington. Sample No. 67.   **1.76**

7. *Hornblende-biotite andesite* cut by the first gas well, south of South Butte.   **1.74**

8. *Biotite rhyolite* from De Witt's Quarry. About 2½ miles N.W. of Sutter City; near the west edge of Sec. 32, T. 16 N., R. 2 E. Sample No. 10. Analysis in Williams' 1929 paper, p. 174.   **1.69**

9. *Hornblende crystals from dacite fragments* close to the base of the middle member of the Rampart formation; valley due south of South Butte.   **1.66**

10. *Hornblende-biotite andesite* from the Twin Peaks dome in the core of the buttes. N.E. corner of Sec. 26, T. 16 N., R. 1 E.   **1.59**

11. *Biotite rhyolite* from the dome forming "777 Hill"; in
    the sedimentary moat on the south side of the buttes.
    Sec. 36, T. 16 N., R. 1 E.   1.55
12. *Hornblende-biotite andesite* from the De Witt volcano.
    About 2 miles W.N.W. of Sutter City. In N.E. ¼ of Sec.
    1, T. 15 N., R. 1 E. "Hill 727." Sample No. 141.   1.50
13. *Hornblende-biotite rhyodacite* from N.E. margin of the
    buttes. N.E. corner of Sec. 5, T. 16 N., R. 2 E. Samples
    Nos. 1 and 68.   1.40

APPENDIX

Petrography

Petrographic notes on the igneous rocks of the Sutter Buttes have already been published along with microdrawings (Williams, 1929). Our purpose here is to summarize and correct some earlier descriptions and to add some new ones, including notes on recently analyzed specimens.

Andesites

All andesites are remarkably similar both in texture and mineral composition. All are strongly porphyritic rocks with a dense pilotaxitic matrix, no matter whether they solidified at the surface or were intruded at depths of as much as 6,000 feet. They vary chiefly in the ratio of hornblende to biotite phenocrysts. In most andesites, hornblende is five to ten times as plentiful as biotite, and in a few it is the only mafic constituent. On the other hand, a few andesites, including the youngest of all, those erupted by the De Witt volcano, contain as much biotite as hornblende. It must be emphasized, however, that hornblende-biotite ratios bear no systematic relation to the ages of the rocks; on the contrary, they may vary considerably within the same body.

Hornblende phenocrysts generally range between 1 and 3 mm. in length, and almost all are euhedral or subhedral. Exceptionally, they may comprise as much as a quarter of the total volume of an andesite, but usually they make up between 5 and 20 percent. No andesite is devoid of them. Their pleochroism is characteristically as follows: X – pale yellow or yellowish green; Y – olive-green or pale bluish green; Z – deep bluish green. Extinction angles, Z Λ c, generally range from 15° to 18°. In a few andesites, however, including some from the De Witt volcano, the hornblende is pleochroic from brown (X) to very deep brown (Y and Z), and the extinction angle is reduced to about 10°. Moreover, as described in Williams' 1929 paper, hornblende phenocrysts in the thin, reddened, reheated crusts of many andesite blocks in the upper member of the Rampart formation are pleochroic from yellow or citrine (X) to orange or deep russet (Y and Z), their extinction angles are commonly reduced to less than 10°, and their birefringence is notably increased. Other hornblende phenocrysts in these reheated crusts are almost completely replaced by granular iron ore.

Biotite phenocrysts generally make up between 2 and 5 percent of the volume of typical andesites, but exceptionally they may be lacking. Most flakes measure between 0.5 and 1.0 mm. in diameter, but they range from minute specks up to 2 mm. in diameter. They are usually pleochroic from pale yellowish green (X) to dark brownish green (Y and Z), and are pseudo-uniaxial. In a few andesites, however, as in many rhyolites, the biotite is strongly zoned from dark olive-green or very dark brown cores to pale yellowish green rims. The reason remains to be determined. In reheated crusts of laharic blocks that contain oxyhornblende, the biotite flakes are reddened and are markedly biaxial (2V to 20°), and some flakes are largely replaced by granular iron ore.

Plagioclase phenocrysts make up between 15 and 30 percent of typical samples. Most measure between 1 mm. and 5 mm. in length, but they range from about 0.2 mm. to 8 mm. long. Almost all exhibit strong oscillatory zoning, with a general increase in soda toward the rims. The rims may consist of oligoclase, but usually the phenocrysts range in composition from medium labradorite to medium andesine.

The pilotaxitic groundmass normally constitutes about half the volume, and consists of microliths of oligoclase, minute grains of magnetite, and interstitial "cryptofelsite." Needles of apatite are abundant, and pores in the groundmass of many andesites contain cristobalite. Sulfides are conspicuously absent.

It must suffice to describe two recently analyzed specimens and some unusual samples of intensely auto-brecciated andesites.

Two samples of *auto-brecciated andesite from the De Witt volcano* have been analyzed (Nos. 5 and 6, see Table 1). These differ from the usual andesites, not only in being auto-brecciated, but in three other particulars: the hornblende and biotite phenocrysts are present in approximately equal amounts, the hornblendes are brownish instead of greenish, and the biotites are much darker than usual. Roughly a third of each sample consists of zoned phenocrysts of medium labradorite to medium andesine. A tenth consists of biotite flakes, with optic angles of about 5°, pleochroic from chestnut-brown to brownish black or almost black. Another tenth consists of hornblende phenocrysts that are pleochroic from light brown (X) to extremely dark brown (Y and Z), with extinction angles, Z Λ c, up to 10°. The matrix of both samples has a pilotaxitic texture composed of minute, stumpy laths and smaller, slender microliths of oligoclase and interstitial "cryptofelsite" carrying needles of apatite, granules of iron ore, and a considerable amount of cristobalite lining irregular pores.

The other recently analyzed andesite (No. 2, see Table 1) comes from the only dike observed in the buttes. Compared with most andesites, this is rich in hornblende and poor in biotite. Zoned phenocrysts of plagioclase ($Ab_{44\text{-}50}$) constitute about 15 percent of the volume. Phenocrysts of hornblende, thinly edged with granular iron ore, make up about a quarter of teh whole. Small flakes of brown biotite make up only about 5 percent of the volume. Accessory constituents include granules of iron ore, needles of apatite, and a few minute grains of diopside. All these minerals lie in a pilotaxitic base of oligoclase microliths and "cryptofelsite."

Reference is made in conclusion to examples of intensely crushed, locally *mylonitized andesites* to be seen in a few places bordering the "central vent-area," especially along its southeast margin. Almost all the plagioclase phenocrysts in these rocks, and many hornblende crystals, are thoroughly fractured, and some biotite flakes are strongly crumpled. Locally, the rocks are riddled by narrow, anastamosing, dark, vein-like mylonitic stringers composed of ultra-comminuted crystals. Noteworthy, though of unknown origin, are occasional minute, well-rounded grains of quartz. These crushed andesites are of cataclastic, rather than protoclastic origin; crushing was not caused by differential movements within semiconsolidated magma but by shattering of already solidified andesite by continued rise of viscous magma beneath a solid roof.

Basic Inclusions

Some andesites adjoining Peace Valley contain a few subangular inclusions of gabbro up to 6 inches across. No other andesites in the buttes are known to carry similar inclusions, though a few are present in the basal, rhyolite-rich member of the Rampart formation on the south side of Peace Valley and on the north side of Kinch Canyon.

These gabbroid inclusions vary from fine- to coarse-grained, equigranular types; mineralogically, most of them consist of green hornblende or actinolite and labradorite with accessory sphene, apatite and iron ore. In some specimens, the amphibole encloses diopside; in others a small amount of biotite is present. Several inclusions have already been described and illustrated, and one has been chemically analyzed (Williams, 1929, pp. 156-157 and plate 18).

The inclusions are either autholiths, i.e. early segregations from the andesitic magmas, as Williams originally thought, or xenoliths torn from the plutonic basement beneath the buttes. Similar gabbroid fragments are present in the basal Rampart beds in Kinch Canyon where they are accompanied by fragments of hornblende-biotite quartz diorite similar to some of the basement-rocks cut by drilling within the buttes. Radiometric dating might solve the problem of the origin of the basic inclusions.

Propylites

Propylites are scattered irregularly in the central part of the buttes but are concentrated chiefly in the northern part. The oldest andesites are generally the most altered, though a few young ones are also propylitized.

Development of propylites was probably caused by hydrothermal solutions charged with carbon dioxide. But the Sutter Buttes propylites, unlike those in many other volcanic regions, are almost completely devoid of sulfides. Their most characteristic feature is the replacement of hornblende phenocrysts by chlorite and calcite, the rims of the pseudomorphs usually being outlined by trains of granular iron ore (see Williams, 1929, plate 19c). Most biotite phenocrysts, on the other hand, remain fresh, though some are replaced by montmorillonite or calcite or both, with separation of granular sphene. Most plagioclase phenocrysts and the pilotaxitic groundmass of the andesites also remain fresh, though the plagioclase phenocrysts in some propylites are almost entirely replaced by calcite, and the groundmass is riddled with veinlets and irregular patches of calcite and quartz.

Rhyolites, Rhyodacites, and Dacites

Many additional chemical analyses are needed before we can be sure about the relative abundance of rhyolites, rhyodacites, and dacites, though rhyolites greatly predominate. Hornblende is usually by far the principal and occasionally the only mafic constituent in andesites, whereas biotite is almost invariably the only mafic constituent in rhyolites. Many lavas and much fragmental debris contain hornblende and biotite in approximately equal amounts, with or without phenocrysts of quartz; these are provisionally grouped as dacites and rhyodacites depending on their content of porphyritic quartz and sanidine. It should be emphasized, however, that samples from the same extrusive dome or intrusive body may vary considerably, even from biotite-rich rhyolite to biotite-free dacite. Dacites and rhyodacites seem to be concentrated close to the southern margin of the andesitic core of the buttes, i.e. south of South Butte, and also close to the northeastern margin, near North Butte.

Rhyolites

Siliceous domes that rise within the sedimentary moat generally consist of porphyritic rhyolite carrying sporadic, rounded phenocrysts of quartz, numerous phenocrysts of sodic plagioclase and a few of sanidine, along with flakes of biotite set in a micro- to crypto-felsitic matrix which, in rare cases, may once have been glassy. The following account begins by describing samples that have been analyzed chemically or dated radiometrically; then we describe a few other rhyolites.

A body of strongly banded, unusually porphyritic rhyolite outcrops along the eastern edge of the andesitic core of the buttes, roughly half a mile north of the old Moore's ranch house and a few hundred yards from the Kellogg ranch house. It is illustrated in Williams' 1929 paper (plate 18, figs. b and c). Approximately half the rock consists of phenocrysts in the following percentages: biotite, 25; quartz, 10; oligoclase and sanidine, each about 6. Sub-parallel flakes of biotite, up to 2 mm. in diameter, are typified by a blotchy distribution of colors. The dominant pleochroism is from pale yellow to deep chestnut-brown, locally to russet. But in most flakes, irregular patches exhibit much more intensive pleochroism, from dark yellow or greenish yellow (X) to extremely dark brown or even to "blackish brown" (Y and Z). These dark patches are usually concentrated in the middle of the flakes, but may be scattered at random throughout. All the biotites are fresh and pseudo-uniaxial. Quartz phenocrysts range up to 3 mm. in diameter; all are corroded and many are deeply embayed bipyramids. Sanidine and oligoclase phenocrysts range up to 4 mm. in length; the former are pseudo-uniaxial and the latter show oscillatory zoning between sodic and calcic oligoclase. The banded matrix consists mainly of "patchy polarizing" microgranular and micrographic intergrowths of quartz and kaolinized feldspar traversed by sub-parallel microliths of sanidine or oligoclase or both. Minute, stumpy prisms of apatite and a few of zircon are scattered throughout, along with occasional radiating clusters of colorless to pale blue tourmaline, the only occurrence of this mineral so far detected in the buttes.

Biotite rhyolite from De Witt's Quarry, near the southern end of the elongate dome a mile or so southeast of the rhyolite just described, has been analyzed chemically (No. 12, see Table 1) and dated radiometrically. This lava was erupted during the explosive phase that built the Rampart formation, so that talus blocks are incorporated in adjacent lahars. It is much less porphyritic than the lava just described but is equally well-banded, and in particular contains no phenocrysts of sanidine and only a few of quartz. Sub-parallel flakes of biotite, strongly pleochroic from golden yellow to russet, make up roughly 8 percent of the bulk, and so do zoned phenocrysts of medium to calcic oligoclase. Prisms of zircon and apatite are lightly disseminated in the groundmass, together with rare, euhedral crystals of primary epidote up to 1 cm. long and even fewer of hornblende. The groundmass consists of microfelsite traversed by swarms of sub-parallel microliths of untwinned and singly twinned feldspar, either sanidine or oligoclase or both. A small amount of cristobalite lines irregular pores.

An intensely auto-brecciated rhyolite from the same dome was collected in the east-west gully a short distance north of De Witt's Quarry. It is made up of angular chips of quartz and kaolinized feldspar, and rare flakes of altered biotite in a silicified matrix stippled with abundant euhedral flakes of kaolinite.

Rhyolite that forms the conspicuous "777 Hill" in the sedimentary moat on the south side of the buttes has been dated radiometrically as 1.55 m.y. old. It is a biotite-rich lava of unusual texture. Cracked and corroded phenocrysts of quartz make up approximately a tenth of the volume; so do flakes of biotite. A few small phenocrysts of oligoclase are also present, but no sanidine was detected. The microfelsitic base contains swarms of isolated and coalescing spheroids, between 0.01 and 0.05 mm. in diameter. Centers of the spheroids seem to be composed of cloudy anhedra of orthoclase and sub-parallel microliths of oligoclase, whereas the rims consist of micro- and crypto-crystalline quartz. The origin of the spheroids is not understood.

An essentially similar rhyolite forms a small dome in the sedimentary moat close to the old Keenan's Ranch House (in the middle of Sec. 15, T. 16 N., R. 1 E.), on the northwest side of the buttes. It is illustrated in plate 18a of Williams' 1929 paper. Restudy of this rock shows that almost half consists of phenocrysts, as follows: zoned calcic to sodic oligoclase (25%); quartz (10%); biotite (8%). Sanidine phenocrysts appear to be absent. The cloudy cryptofelsitic matrix carries abundant microliths of oligoclase, occasional needles of apatite, and a single grain of orthite. Similar biotite-rich rhyolite forms the conspicuous dome on the south side of Peace Valley (N.E. ¼, Sec. 13, T. 16 N., R. 1 E.).

Rhyodacites and Dacites

Rocks that fall under this caption are to be found chiefly along the southern edge of the andesitic core, near South Butte, and on the northeastern side of the buttes. Typically, they lack or contain very few phenocrysts of quartz and sanidine, and carry phenocrysts of hornblende, with or without biotite.

The only rhyodacite analyzed chemically (No. 8) and dated radiometrically comes from the northeastern edge of the buttes. A third of this lava consists of plagioclase phenocrysts up to 6 mm. long, ranging in composition from medium to calcic andesine. Phenocrysts of olive-green hornblende and russet-brown biotite are present in approximately equal amounts, totalling 15 percent of the volume. Only a single phenocryst of quartz was observed. The groundmass consists of cryptofelsite traversed by abundant stumpy microliths of oligoclase.

An essentially similar hornblende-biotite dacite forms a small dome a quarter of a mile south of Dean's Place (Sec. 19, T. 16 N., R. 2 E.). And a short distance north of Dean's Place there is a vitrophyric dacite or rhyodacite that lacks quartz and sanidine phenocrysts, but contains phenocrysts of andesine (20%), biotite (2%), and hornblende (0.5%) in a matrix of colorless glass (R.I.-1.53) charged with microliths of oligoclase and a few prisms of apatite and zircon. Other samples from between Dean's Place and North butte are tentatively classed as phenocryst-poor hornblende-biotite dacite or rhyodacite.

Reference is made in conclusion to the siliceous intrusions along the margin of the andesitic core near South Butte. These are readily recognizable from a distance by reason of their whitish colors, and in hand specimens by their porcelanous appearance. They contrast markedly with the fresh rhyolites and dacites just described from within the sedimentary moat in being much less porphyritic and markedly altered. Most of them have been too much altered hydrothermally to be classified with certainty.

Two samples taken approximately half a mile southeast of South Butte are characteristic. In one, phenocrysts of sodic andesine make up roughly 10 percent of the volume, and flakes of biotite 5 percent. Phenocrysts of quartz and sanidine are lacking. The groundmass is composed of cloudy, kaolinized cryptofelsite containing swarms of slender microliths of singly twinned and untwinned oligoclase or sanidine or both. The other sample has a thoroughly kaolinized and sericitized cryptofelsitic matrix carrying rare phenocrysts of quartz and chloritized biotite, and more numerous phenocrysts of andesine, partly altered to calcite and kaolin.

Another of these enigmatic, altered siliceous rocks with a porcelain-like appearance in hand specimens outcrops at the southwest base of South Butte. Again, there are no phenocrysts of quartz or sanidine. All biotite flakes and much of the cryptofelsitic groundmass have been replaced by mixtures of calcite, sericite, and kaolin.

Rampart Beds

The basal Rampart beds are the richest in reworked magmatic ejecta of rhyolitic and dacitic composition. In the middle member, andesitic debris increases from the bottom upward, and the top member consists almost exclusively of andesitic debris. There are, however, pronounced lateral and vertical variations because the pre-eruption surface was irregular and many Rampart beds were eroded and redeposited during the explosive episode. All three members therefore grade into each other, and because almost all are reworked deposits, their succession reflects only in a general way the actual sequence of erupted magmas.

The initial eruptions are represented by the lowermost beds along the south side of West Butte Pass.

These consist of pumiceous and dense, porphyritic biotite rhyolite rich in phenocrysts of quartz, plagioclase, sanidine, and biotite in a microfelsitic matrix (K/A date—2.4 m.y.). Above them lie waterlaid tuffaceous beds of siliceous, hornblende-biotite andesite.

One of the best sections of the Rampart beds is at the east end of Peace Valley. A sample from near the bottom of the basal member is a biotite rhyolite tuff. It consists of chips of biotite rhyolite mixed with many discrete, broken crystals of plagioclase, a few of sanidine and quartz, and flakes of biotite in a matrix of devitrified glass-dust. Another sample from near the base is made up of angular chips of lava, broken crystals, and devitrified glass-dust, largely altered to clay. Most lithic chips are composed of hornblende- and biotite-poor dacite or rhyodacite similar to lavas composing some of the small domes near North Butte; a few consist of hornblende-biotite andesite.

A white bomb of pumice, collected about 100 feet above the preceding sample, consists of hornblende-biotite dacite or rhyodacite. Phenocrysts constitute roughly a third of the volume, as follows: andesine (20%), green hornblende (6%), and biotite (6%). These crystals, along with a few prisms of apatite and granules of iron ore, lie in a matrix of pumiceous glass with a refractive index of $1.518 \pm .002$, suggesting a silica percentage of approximately 67.

Another pumiceous bomb was collected 250 feet above the one just described and immediately below the blocky andesitic lahars that form the topmost Rampart beds. This bomb differs from the preceding one in carrying more and larger phenocrysts, particularly of hornblende. Phenocrysts make up about 40 percent of the volume, as follows: andesine (20%), green hornblende (17%), yellowish to russet-brown biotite (3%). The matrix of colorless glass (R.I. - $1.518 \pm .002$) is crowded with slender microlites of sanidine or oligoclase or both.

A large pumiceous block from near the base of the Rampart beds in the Morehead "rock-and-gravel pit" at the south edge of the buttes is a pale-colored, siliceous hornblende-biotite andesite or, possibly, dacite. Roughly a third consists of large plagioclase phenocrysts; flakes of biotite make up about 8 percent; and prisms of green hornblende comprise 5 percent. One of two thin-sections carries a single, small phenocryst of quartz, but neither contains sanidine. The matrix is vesicular cryptofelsite, probably representing devitrified glass.

Two samples from the Rampart formation were analyzed chemically, one of rhyodacite (No. 9, see Table 1) and the other of andesite (No. 3). The former came from the basal beds on the northeast side of Kinch Canyon. A third of it consists of plagioclase phenocrysts normally zoned from medium andesine to clacic oligoclase. Sanidine phenocrysts were not detected and only a few small phenocrysts of quartz are present. Sub-parallel flakes of biotite make up about 5 percent of the whole, but minute crystals of oxyhornblende make up only 0.1 percent. The porous cryptofelsitic matrix is charged with microliths of oligoclase or sanidine or both, and many pores are lined with cristobalite.

The other analyzed sample is an andesitic tuff from the local base of the Rampart formation, close to the southwest edge of the buttes, where the Rampart beds rest with strong unconformity on vertical Sutter beds (plate 3b). It comes from well up in the middle member. Half of it consists of broken, zoned crystals of plagioclase (medium labradorite to medium andesine), a tenth is made up of green hornblende, and 1 percent of biotite. These constituents lie in a matrix of colorless, pumiceous glass containing a few minute grains of diopside and needles of apatite. Chemically and mineralogically, the tuff resembles many andesitic lavas forming Pelean domes in the central part of the buttes.

Typical samples of hornblende-biotite andesite from the unaltered cores and reheated crusts of large blocks in the topmost laharic beds have already been described and illustrated (Williams, 1929, plates 17a and b; 19b and d). These came from the steep flanks of the youngest Pelean domes.

Lake Sediments in the Center of the Buttes

The well-bedded, waterlaid sediments in the center of the buttes consist mainly of sand- and gravel-size debris, though many layers, particularly near the top of the sequence, consist of silt- and clay-size particles. The complete absence of ovoid and ropy bombs and lapilli, and of pumiceous fragments and vitric ash, along with the angularity of virtually all the constituents, indicate that the sediments are reworked products of phreatic (steam-blast) and phreatomagmatic eruptions.

The thickest and best exposed section is on the south wall of Bragg's Canyon. The basal part is concealed; hence, we cannot say whether or not it consists mainly of fresh magmatic debris, as does most of the basal member of the Rampart formation. The exposed beds are composed chiefly of fragments of hornblende-biotite andesite, generally fresh but partly propylitized. A sample from near the visible base of the sequence (No. 40, fig. 10) is composed entirely of crystals of zoned plagioclase, flakes of fresh biotite, and thoroughly altered crystals of hornblende, now made up of chlorite, clay, and iron ore. Neither quartz nor sanidine was detected. Presumably, therefore, the sediment was derived from partly propylitized andesites. A second sample (No. 41), collected approximately 100 feet stratigraphically above the first, though slightly coarser-grained, containing angular chips up to 1 cm. across, is essentially similar in composition. It consists mainly of broken crystals of fresh plagioclase, flakes of fresh biotite, and crystals of hornblende, some entirely replaced by chlorite, others by granular iron ore, and still others by mixtures of chlorite, calcite, and clay. The matrix between these crystals and accompanying chips of andesite is a murky, irresolvable clay.

A third sample (No. 42), collected from the lowermost cliffs, approximately 150 feet stratigraphically above the preceding sample, is distinctive by reason of its purplish brown color. It carries angular chips, up to 5 mm. across, of hornblende-rich, biotite-poor andesite, in a matrix of broken crystals of plagioclase, euhedral prisms of hornblende, and flakes of biotite. All the hornblende crystals, which constitute 15 percent of the bulk, are altered; some are entirely replaced by iron ore, but most by a mixture of chlorite, calcite, and iron ore. And though most of the biotite flakes are fresh, many are reddened and rimmed by magnetite. The purplish brown color of the rock appears to be caused by oxidation of secondary magnetite to hematite.

A fourth sample (No. 44), from the ridge-crest close to the eastern margin of the lake-basin, approximately 700 feet stratigraphically above the preceding sample, is also a purplish brown sediment. It is composed largely of chips of hornblende-biotite andesite mingled with discrete crystals of plagioclase and fewer of hornblende and biotite in a subordinate matrix of clay suffused with hematite-dust. Almost all the hornblende crystals, unlike those in the preceding two samples, are fresh and green. In brief, the sediment seems to have been derived from ejecta similar in composition to most of the lavas of the adjacent Pelean domes.

Other samples were collected near the ridge-crest, approximately on the same horizon as the preceding sample, but about 400 yards to the west. These are thinly bedded, sage-green and brownish sands, silts, and clays. Their mineral content varies considerably, even over short distances. Some layers lack rock fragments, being composed of minute chips of quartz and plagioclase and rare flakes of biotite in a matrix of clay; these layers probably represent reworked crystal-vitric ashes of dacitic composition. Other layers lack quartz, consisting chiefly of splinters of plagioclase, flakes of biotite, and more numerous crystals of hornblende largely replaced by chlorite and calcite; these were probably derived from propylitized andesites. Still other samples consist of minute chips of biotite- and hornblende-biotite rhyodacite and rhyolite, along with broken crystals of quartz and plagioclase in a silicifed, clayey matrix reddened with hematite-dust. Texturally and mineralogically, many rock chips resemble the porcelanous, white rhyodacites south of South Butte. Neither here nor elsewhere in the central lake basin did we find lithic fragments of biotite rhyolite similar to the lava that forms many Pelean domes within the sedimentary moat surrounding the andesitic core of the buttes.

Our conclusion is that the waterlaid, dominantly lacustrine sediments consist chiefly of reworked hornblende-biotite andesitic and propylitic ejecta mingled with subordinate rhyolitic, rhyodacitic and dacitic debris.

Reference is made finally to volcanic sediments within the andesitic core of the buttes, but outside the central lake basin. These are well-bedded siltstones and sandstones bordering the trail along Brockman Canyon, to which reference has been made elsewhere. (See "Rampart Beds," above.) Some are composed of chips of plagioclase and fewer of quartz and sanidine, along with many sub-parallel flakes of biotite in a matrix of clay; others contain, in addition to the foregoing minerals, chips of biotite rhyolite or rhyodacite. These siliceous sediments are probably coeval with some of those forming the basal member of the Rampart formation two miles to the south.

Petrology

Analyzed igneous rocks of the Sutter Buttes range in composition from andesite to rhyolite, samples ranging in silica content from 55.42 to 71.20 percent. There is, however, no regular sequence in time, as in many other volcanic fields. All that can be said with certainty is that andesites are concentrated in the core of the buttes, whereas most dacites and rhyolites are concentrated within and beneath the surrounding moat and rampart. We are at a loss to understand why this is so.

According to the classification proposed by Peacock (1931), the igneous rocks of the Sutter Buttes belong to the calc-alkaline series. Their alkali-lime index (i.e. the silica percentage at which the percentage of CaO equals that of Na_2O and K_2O combined) is approximately 60. Roughly coeval lavas of the Newberry volcano in central Oregon and of the Medicine Lake Highlands in northern California, both of which lie east of the Cascade Range, also belong to the calc-alkaline series, their alkali-lime indices approximating 58. On the other hand, roughly coeval Quaternary lavas in the High Cascades of Oregon and northern California belong to the calcic igneous series, their alkali-lime indices ranging from about 61 to 64. Lavas of the same silica content from the Sutter Buttes are poorer in lime and richer in potash than those from the High Cascades. Indeed, andesites from the Sutter Buttes are distinctly richer in potash than most andesites of the same silica-content in other parts of the world. (See fig. 12.)

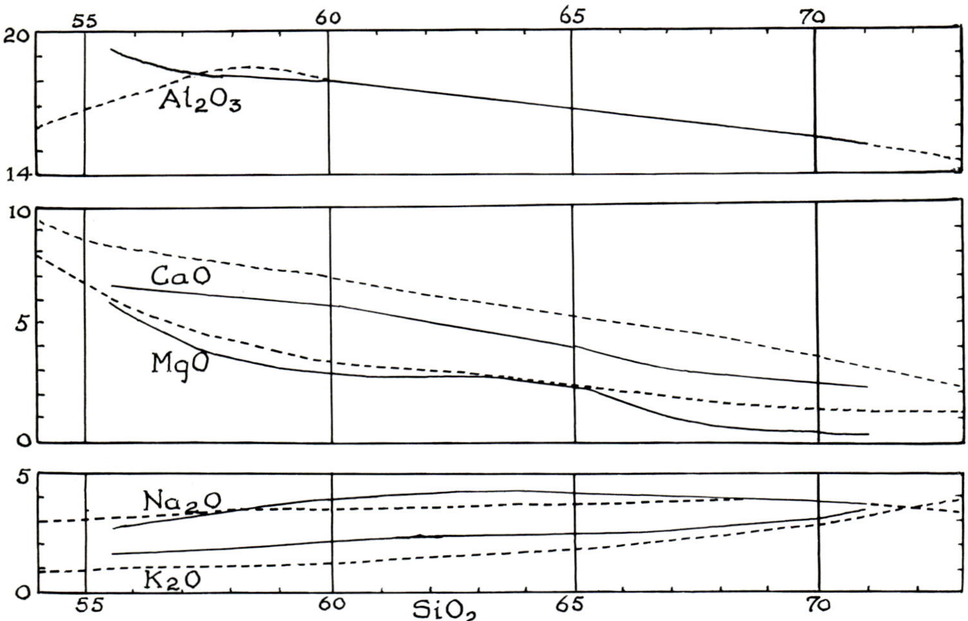

Fig. 12. Variation diagrams comparing Lassen lavas (solid lines) with Sutter Buttes lavas (dashed lines).

Localities of Analyzed Samples

1. *Hornblende-biotite andesite.* Sample 2, 1929. One mile west of the summit of South Butte. Sec. 27, T. 16 N., R. 1 E.
2. *Hornblende-biotite andesite.* Sample 13 c. Columnar dike trending N.W. near the 900-foot contour in creek bed close to the center of Sec. 24, T. 16 N., R. 1 E.
3. *Andesite fragment* from near the base of the middle member of the Rampart formation. Sample 12 c. About 1½ miles E.N.E. of the hamlet of West Butte. Near the center of Sec. 33, T. 16 N., R. 1 E.
4. *Propylite.* Sample 6, 1929. A mile south of Pugh Place, Peace Valley. Sec. 13, T. 16 N., R. 1 E.
5. *Auto-brecciated andesite flow from De Witt's volcano.* Sample 15 c. Sec. 6, T. 15 N., R. 2 E.
6. *Auto-brecciated andesite flow from De Witt's volcano.* Sample 14 c. N.W. ¼, Sec. 1, T. 15 N., R. 1 E.
7. *Hornblende-biotite andesite.* Sample 1, 1929. From the summit of South Butte. Sec. 26, T. 16 N., R. 1 E.
8. *Rhyodacite.* Samples 1 W and 68. From the northeast edge of the buttes. S. edge of Sec. 10, T. 16 N., R. 2 E.
9. *Rhyodacite.* Sample 8 c. Fragment from the basal member of the Rampart formation on N.E. side of Kinch Canyon. E. ½, Sec. 29, T. 16 N., R. 2 E.

10. *Rhyodacite.* Sample 19, 1929. West of North Butte. Sec. 18, T. 16 N., R. 2 E.
11. *Rhyolite.* Sample 13 W, New series; 65 old series. Roughly half a mile S.W. of the Old Keenan Ranch house; N.W. ¼, Sec. 22, T. 16 N., R. 1 E.
12. *Rhyolite.* Sample 10, 1929. From De Witt's Quarry dome, 3 miles N.W. of Sutter City. Sec. 32, T. 16 N., R. 2 E.

As to the depths from which the Sutter Buttes magmas were derived, we can only speculate. Dickinson and Hatherton (1967) pointed out that among circum-Pacific island arcs, the K_2O-content of andesites can be correlated rather closely with increasing depths to underlying Benioff seismic (subduction) zones. Andesitic volcanism, in their opinion, originates "in the mantle where magmas are generated by events associated with earthquake of intermediate focal depths." Many other writers have given support to the idea that the potash content of andesites generally increases inward from continental margins (e.g., Lipman, Prostka, and Christiansen, 1971). Dickinson and Hatherton (1967) also demonstrated that the K_2O content of circum-Pacific andesites is approximately 1.0 percent where depths to the Benioff zone approximate 80 km.; and as the K_2O content increases to about 3.0 percent, depths to the Benioff zone increase in a fairly regular manner to about 300 km.

Dickinson (1970) inferred that Quaternary andesites in the High Cascades of Oregon and northern California originated from a subduction zone at depths of between 110 and 140 km., whereas lavas of the same silica content from the Newberry volcano of central Oregon and the Medicine Lake volcanos of northern California—both of which lie east of the High Cascades—came from depths of 150 and 160 km., respectively, their contents of K_2O at 60% SiO_2 being 1.8 and 2.0. However, the lavas of the Sutter Buttes, though they lie west of the southern prolongation of the High Cascades, and hence are supposed to have originated at shallower depths, seem to have issued from greater depths (approximately 180 km.), since their K_2O content at 60% SiO_2 is 2.14%. The reasons for this apparent anomaly remain obscure. Chemical analyses of Quaternary lavas from the Clear Lake area, which lies about 50 miles W.S.W. of the Sutter Buttes, have been presented by Anderson (1936), Stevenson *et al.* (1971), and Bowman *et al.* (1973). Anderson's variation diagram suggests that if lavas with 60% SiO_2 are present, their K_2O content would approximate 2.0%, suggesting a depth of origin of about 160 km, as at Medicine Lake. Nielson and Stoiber (1973) have shown, however, the danger of using potassium content of andesitic lavas as a basis for estimating the depths of paleosubduction zones and the origins of magmas. (See fig. 13.)

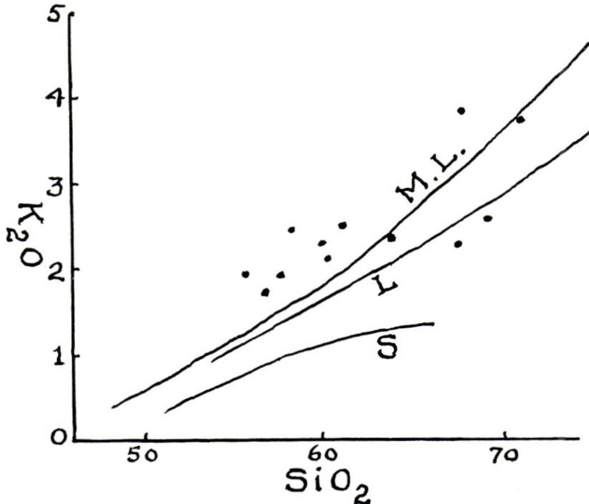

Fig. 13. Plot of K_2O against SiO_2. Lavas from Medicine Lake (ML), Lassen area (L), and Shasta (S) compared with lavas from the Sutter Buttes. Reproduced in part from Smith and Carmichael (1968).

Table 1

IGNEOUS ROCKS FROM THE SUTTER BUTTES

	1	2	3	4	5	6	7	8	9	10	11	12
SiO_2	55.42	56.49	57.48	58.24	59.77	60.24	60.84	63.62	67.70	67.93	68.84	71.20
TiO_2	0.32	0.52	0.53	Tr.	0.62	0.58	0.18	0.08	0.18	0.21	0.14	0.12
Al_2O_3	19.47	16.86	18.07	16.69	17.98	17.78	17.86	16.94	16.43	15.01	16.03	14.96
FeO	2.50	2.78	3.31	1.22	2.92	2.91	1.72	1.83	0.68	0.20	2.74	0.42
Fe_2O_3	3.43	2.47	2.16	3.37	2.41	2.12	2.59	1.87	1.96	1.29	0.12	1.10
MnO	0.13	0.14	0.16	0.25	0.12	0.12	0.16	0.07	0.08	Tr.	0.07	Tr.
CaO	6.96	6.56	6.49	4.62	6.23	6.02	5.24	4.62	3.42	2.03	2.67	2.75
MgO	5.59	6.03	4.13	3.15	2.72	3.26	3.05	2.88	0.88	0.51	0.54	0.62
Na_2O	2.95	3.18	3.32	2.70	3.92	3.78	3.80	4.06	3.87	3.76	4.21	3.90
K_2O	1.88	1.52	1.94	2.46	2.19	2.10	2.48	2.33	2.28	3.93	2.42	3.88
P_2O_5	0.33	0.16	0.16	0.12	0.19	0.17	0.26	0.27	0.08	0.12	0.08	0.15
H_2O-	0.75	1.87	0.48	0.48	0.38	0.25	1.20	0.36	1.27	0.71	0.18	0.20
H_2O+	0.40	1.23	1.82	4.24	0.62	0.69	0.50	1.20	1.23	4.33	1.82	0.55
CO_2	–	Nil	Nil	2.16	Tr.	Tr.	Nil	Nil	Tr.	--	0.37	Nil
Totals	100.13	99.81	100.05	99.70	100.07	100.02	99.88	100.13	100.06	100.03	100.23	99.85

Analyses 1, 4, 7, and 10 by K. Willman; others by W. H. Herdsman

Table 2

TRACE ELEMENTS DETERMINED BY X-RAY FLUORESCENCE

	1	2	6	7	8	9	10	12
Nb	10	<5	–	10	5	5	10	10
Zr	125	125	135	125	130	115	70	115
Y	20	25	20	20	25	20	20	20
Sr	765	750	865	725	835	750	385	475
Rb	75	70	95	85	120	85	185	155
Th	<5	<5	5	5	<5	<5	5	<5
Pb	10	10	20	20	20	10	60	40
Ga	25	15	20	20	20	20	20	15
Zn	60	60	60	55	60	60	30	35
Cu	15	35	20	35	20	30	5	15
Ni	100	165	25	95	45	45	10	10
V	110	120	90	110	65	125	–	5
Cr	260	420	75	250	120	135	5	15
Ba	1135	1110	1300	1050	1680	1060	1330	1905
Ce	50	55	50	45	45	60	25	35
Nd	20	25	20	20	20	25	10	15
Pr	5	10	15	10	5	20	10	5
La	15	25	20	25	25	25	5	10
Cl	120	140	170	220	60	350	90	–
S	30	70	40	130	70	100	50	10

Concentrations in parts per million. Analyses by Robert Jack.

Trace Elements. Estimates of the content of trace elements in the Sutter Buttes lavas were made for us at Berkeley by Robert Jack, using X-ray fluorescence methods; Lowder (1970) used the same apparatus in his study of lavas from the Talasea region of New Britain. Steinborn (1972) used instrumental neutron activation techniques to determine trace-element abundances in Quaternary High Cascade lavas from Oregon and northern California. More complete determinations were made by Taylor and White (1966) on Quaternary andesites from New Zealand and Japan. Results of the foregoing studies lead to the following conclusions.

Andesites from the Sutter Buttes are much richer in Ba (average 1,300 p.p.m. vs. 270-480) and in Rb (95 vs. 48) than approximately coeval High Cascade lavas of the same silica-content. They are also richer in K_2O (2.2% vs. 1.4%), but slightly poorer in Na_2O (3.8% vs. 4.0%). On the other hand, andesites from the Sutter Buttes are somewhat poorer in Ti, V, Cu, and Zn.

Compared with Quaternary andesites from New Zealand and Japan, those from the Sutter Buttes are much richer in Ba and Sr, slightly richer in Zr, Rb, Ni, and Cr, but again slightly poorer in Cu and V. Other trace elements are present in essentially the same amounts.

Compared with andesites from the Talasea region of New Britain, those from the Sutter Buttes are again much richer in Ba and Rb, but poorer in Cu and Zn.

Only one rhyolite from the Sutter Buttes was examined for its trace elements; this is much richer in Ba and Sr but poorer in Rb and Y than the Quaternary rhyolites of the Clear Lake area.

The unusually high content of Ba and Sr in the Sutter Buttes lavas remains to be explained.

References

ANDERSON, C. A.
  1936. Volcanic history of the Clear Lake area, California. Bull. Geol. Soc. Amer., 47:629-664.
BOWMAN, H. R., F. ASARO and I. PERLMAN
  1973. On the uniformity of composition in obsidians and evidence for magmatic mixing. Jour. Geol., 81:312-327.
CHAPMAN, R. H.
  1966. The gravity field in northern California. Calif. Div. Mines and Geol., Bull. 190:395-405.
DALRYMPLE, G. B.
  1964. Cenozoic chronology of the Sierra Nevada. Univ. Calif. Pubs. Geol. Sci., 47:1-41.
DICKERSON, R. E.
  1913. Fauna of the Eocene at Marysville Buttes. *Ibid.*, 7:257-298.
DICKINSON, W. R., and R. HATHERTON
  1967. Andesitic volcanism and seismicity around the Pacific. Science, 157:801-803.
DICKINSON, W. R.
  1970. Relations of andesites, granites, and derivative sandstones to arch-trench tectonics. Rev. Geophys. and Space Physics, 8:813-860.
GARRISON, L. E.
  1962. The Marysville (Sutter) Buttes, California. Calif. Div. Mines and Geol., Bull. 181:69-72.
GILBERT, G. K.
  1877. Report on the geology of the Henry Mountains. U.S. Geogr. Geol. Survey of the Rocky Mts. region. 170 pp.
GRANTZ, ARTHUR, and ISIDORE ZEITZ
  1960. Possible significance of broad magnetic highs over belts of moderately deformed sedimentary rocks in Alaska and California. U.S. Geol. Survey Prof. Paper 40-B:342-347.
GRISCOM, ANDREW
  1966. Magnetic data and regional structure in Northern California. Calif. Div. Mines and Geol., Bull. 190:407-417.
HACKEL, OTTO
  1966. Summary of the geology of the Great Valley. *Ibid.*, 190:217-238.
HAWLEY, A. S.
  1962. Wild Goose gas field. *Ibid.,* Bull. 181:109-112.
HEDGE, C. E. and Z. E. PETERMAN
  1967. Sr^{87}/Sr^{86} of circum-Pacific andesites. Geol. Soc. Amer. Abstr., Part 1, p. 96.
HUNT, C. B.
  1953. Geology and geography of the Henry Mts. region, Utah. U.S. Geol. Survey Prof. Paper 228.
HUNT, C. B., and A. C. WATERS
  1958. Structural and igneous geology of the La Sal Mts., Utah. *Ibid.*, 294.
JOHNSON, A. M., and D. D. POLLARD
  1973. Mechanics of growth of some laccolithic intrusions in the Henry Mts., Utah. Tectonophysics, 18:261-354.
JOHNSON, H. R.
  1943. Marysville Buttes (Sutter Buttes) gas field. Calif. Div. Mines and Geol., Bull. 118:610-615.
LACHENBRUCH, M. C.
  1962. Geology of the west side of the Sacramento Valley. *Ibid.*, Bull. 181:53-66.
LINDGREN, WALDEMAR
  1895. U.S. Geol. Survey, Marysville Folio, No. 17.
LIPMAN, P. W., H. J. PROSTKA and R. L. CHRISTIANSEN
  1971. Evolving subduction zones in the western United States. Science, 174:821-825.

LOWDER, G. G., and I. S. E. CARMICHAEL
    1970. The volcanoes and caldera of Talasea, New Britain. Bull. Geol. Soc. Amer., 81:17-38.
MAY, J. C., and R. L. HEWITT
    1948. The Basement Complex in well samples from the Sacramento and San Joaquin valleys,
           California. Calif. Jour. Mines and Geol., 44:129-158.
MINAKAMI, T., T. ISHIKAWA and K. YAGI
    1951. The 1944 eruption of Volcano Usu in Hokkaido, Japan. Bull. Volcanologique, 11:45-157.
NIELSON, D. R., and R. E. STOIBER
    1973. Relationships of potassium content in andesitic lavas and depth to the seismic zone. Jour.
           Geophys. Research, 78:6887-6892.
PEACOCK, M. A.
    1931. Classification of igneous rock series. Jour. Geol., 39:54-67.
PERRET, F. A.
    1937. The eruption of Mt. Pelee 1929-1932. Carnegie Instit. Washington, Public, No. 458.
REPENNING, C. A.
    1960. Geologic summary of the Central Valley of California. U.S. Geol. Survey T.E.I. reprt
           769. (Paleographic maps reproduced in Hackel, 1966).
RUSSELL, R. D., and V. L. VANDERHOOF
    1931. A vertebrate fauna from a new Pliocene formation in northern California. Univ. Calif.
           Pubs. Bull. Dept. Geol. Sci., 20:11-21.
SMITH, A. L., and I. S. E. CARMICHAEL
    1968. Quaternary lavas from the southern Cascades. Contr. Mineral. and Petrol., 19:212-238.
STEINBORN, T. L.
    1972. Trace element geochemistry of several volcanic centers in the High Cascades. Unpublished
           Master's thesis, Univ. Oregon, Eugene.
STEVENSON, D. P., F. H. STROSS and R. F. HEIZER
    1971. An evaluation of X-ray fluorescence analysis as a method for correlating obsidian artifacts
           with source location. Archaeology, 13:17-25.
TAYLOR, G. A.
    1958. The 1951 eruption of Mount Lamington, Papua. Australian Bur. Min. Resources, Geol.
           and Geophys., Bull. 38.
TAYLOR, S. R. and A. J. WHITE
    1966. Trace element abundances in andesites. Bull. Volcanologique, 29:177-193.
THAMER, D. H.
    1961. Report of the Annual Field Trip of Geol. Soc. Sacramento, pp. 36-41.
WILLIAMS, HOWEL
    1929. Geology of the Marysville Buttes, California. Univ. Calif. Pubs. Bull. Dept. Geol. Sci.,
           18:103-220. (This paper contains all pertinent references up to 1929.)
    1935. Newberry volcano of central Oregon. Bull. Geol. Soc., Amer., 46:253-304.

PLATES

1. Sutter Buttes from the air; view **looking north**. Note the rugged **andesitic** core, the discontinuous moat cut in upturned sediments, and the encircling rampart composed mainly of waterlaid volcanic debris. Note also the meandering Sacramento River, the foothills of the Sierra Nevada, and, in the distance, the snow-capped Lassen Peak and Mount Shasta volcanoes. Photo by U.S. Air Force, July, 1968.

2a. Sutter Buttes from the southeast. The long, gentle slopes of "The Rampart" almost entirely conceal the upturned sedimentary beds in "The Moat." The "Castellated Core" includes, from left to right, the andesitic domes of South Butte, Twin Buttes, and North Butte. The flattish skyline near the middle of the picture is occupied by volcanic lake-sediments. Photo by Joachim Hampel.

2b. Sutter Buttes from the northeast. North Butte dome in the middle distance, to the right of the center; Twin Peaks and South Butte domes in the background. The east-west valley transecting the buttes includes Bragg's Canyon. The gentle, outward-dipping slopes are part of the Rampart, and the two small, wooded hills in the meadows (right foreground) are parts of a rhyodacite dome. Photo by Pacific Resources, Inc., Oakland Airport.

3a. *View in the sedimentary moat.* Butte Gravels (Eocene) standing vertically. South of South Butte, near the road through West Butte Pass.

3b. *View in the sedimentary moat.* Near southwest edge of the buttes. Gently dipping Rampart beds to left of the drilling rig; beneath them, from left ro right, vertical and steeply dipping Sutter beds, Butte Gravels, Capay beds, and white Kione sands, the last occupying the gully. View looking north from near center of Sec. 33, T. 16 N, R. 1 E.

4a. Dacite dome of Syowa Sinzan, Hokkaido punched through uplifted volcanic sediments of the "Roof Mountain."

4b. Andesite dome of North Butte. Photo by Joachim Hampel.

5a. Rhyolite dome on De Witt's Ranch; southeast side of the buttes. View looking north. Sutter beds exposed near the edge of the dome, bordering the trail. The hill to the right consists of Rampart beds.

5b. Well-bedded, fluviatile and lacustrine volcanic sediments on the south side of Bragg's Canyon.

6a. Typical appearance of Rampart beds on the northwest side of the buttes, west of Keenan's Ranch (Sec. 16, T. 16 N., R. 1 E.).

6b. Rampart beds on the northwest side of the buttes. Whitish beds at the base belong to the basal, rhyolite-rich member; overlying beds of the middle member consist chiefly of waterlaid andesitic debris (Sec. 14, T. 16 N., R. 1 E.).

6c. Andesite boulders in the youngest laharic deposits. Approximately 2 miles from their source on the dome of North Butte.